Janusz Kacprzyk, Frederick E. Petry, and Adnan Yazici (Eds.)

Uncertainty Approaches for Spatial Data Modeling and Processing

Studies in Computational Intelligence, Volume 271

Editor-in-Chief
Prof. Janusz Kacprzyk
Systems Research Institute
Polish Academy of Sciences
ul. Newelska 6
01-447 Warsaw
Poland
E-mail: kacprzyk@ibspan.waw.pl

Further volumes of this series can be found on our homepage: springer.com

Vol. 250. Raymond Chiong and Sandeep Dhakal (Eds.)
Natural Intelligence for Scheduling, Planning and Packing Problems, 2009
ISBN 978-3-642-04038-2

Vol. 251. Zbigniew W. Ras and William Ribarsky (Eds.)
Advances in Information and Intelligent Systems, 2009
ISBN 978-3-642-04140-2

Vol. 252. Ngoc Thanh Nguyen and Edward Szczerbicki (Eds.)
Intelligent Systems for Knowledge Management, 2009
ISBN 978-3-642-04169-3

Vol. 253. Roger Lee and Naohiro Ishii (Eds.)
Software Engineering Research, Management and Applications 2009, 2009
ISBN 978-3-642-05440-2

Vol. 254. Kyandoghere Kyamakya, Wolfgang A. Halang, Herwig Unger, Jean Chamberlain Chedjou, Nikolai F. Rulkov, and Zhong Li (Eds.)
Recent Advances in Nonlinear Dynamics and Synchronization, 2009
ISBN 978-3-642-04226-3

Vol. 255. Catarina Silva and Bernardete Ribeiro
Inductive Inference for Large Scale Text Classification, 2009
ISBN 978-3-642-04532-5

Vol. 256. Patricia Melin, Janusz Kacprzyk, and Witold Pedrycz (Eds.)
Bio-inspired Hybrid Intelligent Systems for Image Analysis and Pattern Recognition, 2009
ISBN 978-3-642-04515-8

Vol. 257. Oscar Castillo, Witold Pedrycz, and Janusz Kacprzyk (Eds.)
Evolutionary Design of Intelligent Systems in Modeling, Simulation and Control, 2009
ISBN 978-3-642-04513-4

Vol. 258. Leonardo Franco, David A. Elizondo, and José M. Jerez (Eds.)
Constructive Neural Networks, 2009
ISBN 978-3-642-04511-0

Vol. 259. Kasthurirangan Gopalakrishnan, Halil Ceylan, and Nii O. Attoh-Okine (Eds.)
Intelligent and Soft Computing in Infrastructure Systems Engineering, 2009
ISBN 978-3-642-04585-1

Vol. 260. Edward Szczerbicki and Ngoc Thanh Nguyen (Eds.)
Smart Information and Knowledge Management, 2009
ISBN 978-3-642-04583-7

Vol. 261. Nadia Nedjah, Leandro dos Santos Coelho, and Luiza de Macedo de Mourelle (Eds.)
Multi-Objective Swarm Intelligent Systems, 2009
ISBN 978-3-642-05164-7

Vol. 262. Jacek Koronacki, Zbigniew W. Ras, Slawomir T. Wierzchon, and Janusz Kacprzyk (Eds.)
Advances in Machine Learning I, 2009
ISBN 978-3-642-05176-0

Vol. 263. Jacek Koronacki, Zbigniew W. Ras, Slawomir T. Wierzchon, and Janusz Kacprzyk (Eds.)
Advances in Machine Learning II, 2009
ISBN 978-3-642-05178-4

Vol. 264. Olivier Sigaud and Jan Peters (Eds.)
From Motor Learning to Interaction Learning in Robots, 2009
ISBN 978-3-642-05180-7

Vol. 265. Zbigniew W. Ras and Li-Shiang Tsay (Eds.)
Advances in Intelligent Information Systems, 2009
ISBN 978-3-642-05182-1

Vol. 266. Akitoshi Hanazawa, Tsutom Miki, and Keiichi Horio (Eds.)
Brain-Inspired Information Technology, 2009
ISBN 978-3-642-04024-5

Vol. 267. Ivan Zelinka, Sergej Celikovský, Hendrik Richter, and Guanrong Chen (Eds.)
Evolutionary Algorithms and Chaotic Systems, 2009
ISBN 978-3-642-10706-1

Vol. 268. Johann M.Ph. Schumann and Yan Liu (Eds.)
Applications of Neural Networks in High Assurance Systems, 2009
ISBN 978-3-642-10689-7

Vol. 269. Francisco Fernández de de Vega and Erick Cantú-Paz (Eds.)
Parallel and Distributed Computational Intelligence, 2010
ISBN 978-3-642-10674-3

Vol. 270. Zong Woo Geem
Recent Advances in Harmony Search Algorithm, 2010
ISBN 978-3-642-04316-1

Vol. 271. Janusz Kacprzyk, Frederick E. Petry, and Adnan Yazici (Eds.)
Uncertainty Approaches for Spatial Data Modeling and Processing, 2010
ISBN 978-3-642-10662-0

Janusz Kacprzyk, Frederick E. Petry, and Adnan Yazici
(Eds.)

Uncertainty Approaches for Spatial Data Modeling and Processing

A Decision Support Perspective

Springer

Prof. Janusz Kacprzyk
Systems Research Institute
Polish Academy of Sciences
Ul. Newelska 6
01-447 Warszawa
Poland
E-mail: kacprzyk@ibspan.waw.pl

Prof. Adnan Yazici
Multimedia Database Laboratory
Department of Computer Engineering
Middle East Technical University
06531, Ankara
Turkey
Email: yazici@ceng.metu.edu.tr

Dr. Frederick E. Petry
Computer Scientist, Section 7440.5
Naval Research Laboratory
Stennis Space Center, MS 39529
USA
Email: fpetry@nrlssc.navy.mil

ISBN 978-3-642-26252-4 e-ISBN 978-3-642-10663-7

DOI 10.1007/978-3-642-10663-7

Studies in Computational Intelligence ISSN 1860-949X

Softcover reprint of the hardcover 1st edition 2010

Typeset & Cover Design: Scientific Publishing Services Pvt. Ltd., Chennai, India.

Printed in acid-free paper

9 8 7 6 5 4 3 2 1

springer.com

To the memory of Professor Ashley Morris

Foreword

We are facing an immense growth of digital data and information resources, both in terms of size, complexity, modalities and intrusiveness. Almost every aspect of our existence is being digitally captured. This is exemplified by the omnipresent existence of all kinds of data storage, far beyond those stored in traditional relational databases. The spectrum of data being digitally stored runs from multimedia data repositories to your purchases in most stores. Every tweet that you broadcast is captured for posterity. Needless to say this situation posses new research opportunities, challenges and problems in the ways we store, manipulate, search, and - in general - make use of such data and information.

Attempts to cope with these problems have been emerging all over the world with thousands of people devoted to developing tools and techniques to deal with this new area of research. One of the prominent scholars and researchers in this field was the late Professor Ashley Morris who died suddenly and tragically at a young age. Ashley's career begun in industry, where he specialized in databases. He then joined the internationally recognized team at Tulane University in New Orleans working on databases. He was a doctoral student of Professor Fred Petry, one of the leading researchers in the field. As a student, Ashley began an active research program. He soon established himself as a well-known specialist in the areas of uncertainty in databases and GIS systems. He was especially interested in aspects of decision analysis. He received his Ph.D. in Computer Science from Tulane University and then joined the Department of Computer Sciences first at the University of Idaho in Moscow, Idaho, and then at DePaul University in Chicago, Illinois.

It is somewhat uncomfortable to have to commemorate a person taken from us in the prime of his life. In addition to being an excellent researcher and scholar, Ashley was also a warm and friendly human being. On the occasions when I met him at various conferences, I always enjoyed speaking with him. He was a large and happy presence. His enthusiasm and hard work amplified in recent years as he published more and more significant papers. He began attending conferences around the world. Among his friends and colleagues, Ashley will certainly be remembered for many years to come, not only for his scientific accomplishments but also for his extraordinary human qualities. He will also be remembered by people who did not directly know him via his publications and other professional accomplishments.

The international database community should be grateful to Professors Kacprzyk, Petry and Yazici for their initiative in publishing this commemorative volume dedicated to the memory of Professor Ashley Morris. The editors have succeeded in gathering many interesting papers by top people in the field. These authors have paid a final tribute to Ashley in the best way they could, by writing a papers on a topic related to what Ashley enjoyed so much.

New York, City, NY
July 2009

Ronald R. Yager

Preface

This volume is dedicated to the memory of Professor Ashley Morris who passed away some two years ago. Ashley was a close friend of all of us, the editors of this volume, and was also a Ph.D. student of one of us. We all had a chance to not only fully appreciate, and be inspired by his contributions, which have had a considerable impact on the entire research community. Due to our personal relations with Ashley, we also had an opportunity to get familiar with his deep thinking about the areas of his expertise and interests. Ashley has been involved since the very beginning of his professional career in database research and practice. Notably, he introduced first some novel solution in database management systems that could handle imprecise and uncertain data, and flexible queries based on imprecisely specified user interests. He proposed to use for that purpose fuzzy logic as an effective and efficient tool. Later the interests of Ashley moved to ways of how to represent and manipulate more complicated databases involving spatial or temporal objects. In this research he discovered and pursued the power of Geographic Information Systems (GISs).

These two main lines of Ashley's research interests and contributions are reflected in the composition of this volume. Basically, we collected some significant papers by well known researchers and scholars on the above mentioned topics. The particular contributions will now be briefly summarized to help the reader get a view of the topics covered and the contents of the particular contributions.

Part I, "Decision Support, OLAP, Data Fusion and GIS", contains contributions that are related to the areas that are the main subject of this volume, and in which various representations and ways of processing of uncertain, imprecise, incomplete, and imperfect information and knowledge play and considerable role.

Ashley Morris, Piotr Jankowski, Brian S. Bourgeois and Frederick E. Petry ("Decision Support Classification of Geospatial and Regular Objects Using Rough and Fuzzy Sets") explore how to use some tools and techniques based on fuzzy sets and rough sets to better classify and categorize both regular and geospatial objects which are characterized by uncertain, fuzzy, or indeterminate boundaries.

Guy De Tré, Jozo Dujmović and Nico Van de Weghe ("Supporting spatial decision making by means of suitability maps") show how the method of Logic Scoring of Preference (LSP) helps overcome the inadequacies of the traditional Boolean logic based methods used in Spatial Decision Support Systems (SDSS) that are to be used for supporting the solution of semi-structured spatial decision making problems. The authors propose the use of LSP to produce so-called dynamic, geographic suitability maps (S-maps) which provide specialised information on the suitability degree of selected geographic regions for a specific purpose. The approach is based on soft computing and many-valued logics.

Vincent B. Robinson ("Exploring the Sensitivity of Fuzzy Decision Models to Landscape Information Inputs in a Spatially Explicit Individual-Based Ecological Model") deals with how to incorporate fuzzy logic into spatially explicit, individual-based ecological models of dispersal, extending previous works by the author. A prototypical model of small mammal dispersal behavior is used to demonstrate how the fuzzy control of dispersal agents could be implemented, and how the Extensible Component Objects for Constructing Observable Simulation Models (ECO-COSM) system could be loosely coupled with geographic information system (GIS) database for spatially explicit ecological simulation modeling of individual behavior. From a geocomputational management perspective, an animal agent must be able to query the state of relevant GIS layers within its local perceptual range and use that information to make decisions regarding its movement behavior. This is handled by the Probe mechanism. By obtaining Probes from relevant Probeable landscape layers, an agent can acquire a perceptual inventory of its world. A fuzzy formulation of the dispersal model is used to implement four different fuzzy decision models, and a preliminary analysis of the sensitivity of results to variations in selected parameters of the fuzzy model is given.

Anne Laurent ("Fuzzy Multidimensional Databases") proposes to enhance the multidimensional data model to handle fuzziness by extending her original conceptual approach, notably applied in the context of data mining. The novel model proposed provides the way to apply OLAP (On-Line Analytical Processing) methods on fuzzy multidimensional databases, leading to Fuzzy-OLAP Mining.

Panagiotis Chountas, Ermir Rogova and Krassimir T. Atanassov ("Expressing Hierarchical Preferences in OLAP Queries") propose how to introduce the concept of hierarchical preferences into the OLAP query answering requirements for a knowledge based treatment of user requests. The authors introduce an automatic analysis of queries according to concepts defined as part of knowledge based hierarchy to guide the query answering as part of a data-warehouse environment. As a formal tool, hierarchical intuitionistic fuzzy sets, H-IFS, are used and an ad-hoc utility build on top of current OLAP tools like Oracle10g is proposed that allows to enhance the query capabilities of by providing better and knowledgeable answers to user's requests.

Gloria Bordogna, Marco Pagani and Gabriella Pasi ("Imperfect Multisource Spatial data Fusion based on a Local Consensual Dynamics") consider strategies for multisource spatial data fusion which have generally to cope with distinct kinds of uncertainty, related to both the trust of the information source, the imperfection of spatial data, and the vagueness of the fusion strategy itself. The authors propose a parametric fusion method modeling consensual dynamics. Several fusion strategies are discussed ranging from a risk-taking to a risk-adverse attitude. The fusion is quantifier driven, reflecting the concept of a fuzzy majority, and is implemented via a novel generalized OWA operator with importance. The obtained fused map is determined in each location by a distinct majority of the sources that are locally in agreement.

Part II, "Database Querying, Spatial and Temporal Databases", is concerned with various approaches to the extensions of traditional databases, notably to spatial and temporal ones, and to new developments in more human consistent, and effective and efficient flexible querying tools and techniques, in particular based on fuzzy logic.

Aziz Sözer, Adnan Yazıcı, Halit Oguztüzün and Fred E. Petry ("Querying Fuzzy Spatiotemporal Databases: Implementation Issues") discuss the modeling and querying of spatiotemporal data, in particular fuzzy and complex spatial objects representing geographic entities and relations. These are very important topics that are crucial for many applications in geographic information systems. As a follow up to their recent article, the authors focus on the issues that arise from implementing this approach. To be more specific, a case study of the implementation of a meteorological database is considered that combines an object-oriented database with a knowledge base.

Sławomir Zadrożny and Janusz Kacprzyk ("Bipolar queries: a way to deal with mandatory and optional conditions in database querying") discuss an approach to bipolar queries starting with the original idea of Lacroix and Lavency. The authors point out two main lines of research: the one focusing on a formal representation within some well established theories and the analysis of a meaningful combinations of multiple conditions, and the one concerned mainly with the study of how to aggregate mandatory (negative, or required) and optional (positive, or desired) conditions. Emphasis is on the second line of reasoning and some relations with other approaches are shown, notably with Chomicki's queries with preferences, and Yager's works in multicriteria decision making. The authors propose a fuzzy counterpart of a new relational algebra operator winnow and show how a bipolar query can be represented via the select and winnow operators.

Patrick Bosc and Olivier Pivert ("On Some Uses of a Stratified Divisor in an Ordinal Framework") discuss how to take into account preferences in division like database queries. Ordinal preferences are employed which are not too demanding for a casual user. Moreover, the type of query considered is inspired by the division operator and some of its variations where preferences apply only to the divisor. The division aims at retrieving the elements associated with a specified set of values and in a similar spirit, while the anti-division looks for elements which are associated with none of the values of a given set. Queries mixing those two aspects are discussed. Some formal properties are analyzed. The implementation of such queries using a regular database management system is considered.

Gregory Vert and S.S. Iyengar ("Integration of Fuzzy ERD Modeling to the Management of Global Contextual Data") discuss the idiosyncrasies of managing the new paradigm of global contextual data, sets of context data and super sets of context data, and introduce some of the basic ideas behind contexts, extending their previous works. The authors then develop a model for the management of aggregated sets of contextual data and propose methods for dealing with the selection and retrieval of context data that is inherently ambiguous about what to retrieve for a given query. Because contexts are characterized by four dimensions: time, space, impact and similarity, they are inherently complicated to deal with. An original model for spatial-temporal management is presented and then analyzed.

Mohamed Ali Ben Hassine, José Galindo and Habib Ounelli ("Repercussions of Fuzzy Databases Migration on Programs") consider the problem of a smooth migration towards the fuzzy database technology that makes it possible to deal with uncertain or incomplete information showing also the efficiency of processing fuzzy queries, focusing on the impact on programs. Clearly, the integration of the fuzzy databases advantages (flexible querying, handling imprecise data, fuzzy data mining, ..) should minimize the transformation costs. However, the migration of applications or databases in corporate applications arises from changes in business demands or technology challenges, and should improve operational efficiency or manage risk, data migration outage, and performance.

We wish to thank all the contributors for their excellent works that are both related to Ashley's research interests and his main original contributions to the field. We also wish to thank Professor Ronald R. Yager, who has known Ashley for years and has a deep knowledge on the areas covered, for having written the foreword to this volume. We think that a volume like this, containing great contributions by well known people, will be the best tribute to Ashley, a person so much devoted to study and research. We hope that this piece of work can also be a token of appreciation of the entire research community for his long time work and devotion.

August 2009

Janusz Kacprzyk
Frederick E. Petry
Adnan Yazici

Contents

Part I: Decision Support, OLAP, Data Fusion and GIS

Decision Support Classification of Geospatial and Regular Objects Using Rough and Fuzzy Sets 3
Ashley Morris, Piotr Jankowski, Brian S. Bourgeois, Frederick E. Petry

Supporting Spatial Decision Making by Means of Suitability Maps 9
Guy De Tré, Jozo Dujmović, Nico Van de Weghe

Exploring the Sensitivity of Fuzzy Decision Models to Landscape Information Inputs in a Spatially Explicit Individual-Based Ecological Model 29
Vincent B. Robinson

Fuzzy Multidimensional Databases 43
Anne Laurent

Expressing Hierarchical Preferences in OLAP Queries 61
Panagiotis Chountas, Ermir Rogova, Krassimir Atanassov

Imperfect Multisource Spatial Data Fusion Based on a Local Consensual Dynamics 79
Gloria Bordogna, Marco Pagani, Gabriella Pasi

Part II: Database Querying, Spatial and Temporal Databases

Querying Fuzzy Spatiotemporal Databases: Implementation Issues 97
Aziz Sözer, Adnan Yazıcı, Halit Oğuztüzün, Frederick E. Petry

Bipolar Queries: A Way to Deal with Mandatory and Optional Conditions in Database Querying ... 117
Sławomir Zadrożny, Janusz Kacprzyk

On Some Uses of a Stratified Divisor in an Ordinal Framework ... 133
Patrick Bosc, Olivier Pivert

Integration of Fuzzy ERD Modeling to the Management of Global Contextual Data ... 155
Gregory Vert, S.S. Iyengar

Repercussions of Fuzzy Databases Migration on Programs ... 175
Mohamed Ali Ben Hassine, José Galindo, Habib Ounelli

Author Index ... 195

Part I

Decision Support, OLAP, Data Fusion and GIS

Decision Support Classification of Geospatial and Regular Objects Using Rough and Fuzzy Sets

Ashley Morris, Piotr Jankowski, Brian S. Bourgeois, and Frederick E. Petry

Abstract. In Geospatial Databases, there is often the need to store and manipulate objects with uncertain, fuzzy, or indeterminate boundaries. Both fuzzy sets and rough sets have been used with success in this undertaking. In this paper we explore how we can use both of these techniques to better classify and categorize both regular objects and geospatial objects.

Keywords: fuzzy sets, rough sets, spatial databases, geospatial databases, uncertainty, OWA.

Introduction

Fuzzy set theory has been used frequently in helping geospatial analysts classify objects [1]. These analysts and decision makers often encounter remotely sensed data from satellites, aerial photography, or hyperspectral sensors. With this data, they typically must classify the ground cover into one of several pre-defined categories. Many times, there is no clear demarcation line where one category would begin and another would end.

For example, the classic question asked about uncertain boundaries in spatial databases is "where does a forest begin?" Exactly what is the tree density where something can be considered forest? 80 trees per acre are definitely a forest, but what about 20? Also, what if the 20 trees are old growth forest, so it is likely they are much larger in diameter than typical trees?

Frequently, traditional Geospatial Information System (GIS) software forces the GIS analyst to make a crisp boundary between categories like grassland and forest. Fuzzy set theory has allowed advanced GIS to create a gradient ecotone between these crisp categorizations.

Rough set theory is often generalized as the "egg yolk" approach, where the core (yolk) has complete membership in the target set, and the "white" symbolizes all areas of partial membership. [2]

Ashley Morris
DePaul University, School of CTI, Chicago, USA

Piotr Jankowski
San Diego State University, Department of Geography, USA

Brian S. Bourgeois and Frederick E. Petry
Naval Research Laboratory, Stennis Space Center, USA

J. Kacprzyk et al. (Eds.): Uncer. Appro. for Spatial Data Model. and Process., SCI 271, pp. 3–8.
springerlink.com

By using the best attributes of both of these techniques, we believe that we can better categorize geospatial areas than is typically done today.

Similarity-Based Fuzzy Relational Databases

The similarity-based fuzzy model of a relational database, proposed first in [3], is actually a formal generalization of the ordinary relational database model. The model, based on the max-min composition of a similarity relation utilized as the extension of the classical identity relation coming from the theory of crisp sets [4].

The most distinctive qualities of the fuzzy relational database are: (1) allowing non-atomic domain values, when characterizing particular attributes of a single entity and (2) generation of equivalence classes (affecting such basic properties of relational database as the removal of redundant tuples) with the support of similarity relation [5] applied in the place of traditional identity relation.

An attribute value is allowed to be a subset of the whole base set of attribute values describing a particular domain. Any member of the power set of accepted domain values can be inserted as an attribute descriptor except the null set. For an attribute D_j when $|D_j|=n$, then the power set has 2^n values and we will denote this as 2^{D_j} for each domain.

A fuzzy database relation, R is a subset of the cross product of all power sets of its constituent attributes $2^{D_1}\times 2^{D_2}\times\ldots\times 2^{D_m}$

$$R \subseteq 2^{D_1}\times 2^{D_2}\times\ldots\times 2^{D_m}$$

An arbitrary fuzzy tuple t_i is any member of R and has the form $t_i=(d_{i1},d_{i2},\ldots,d_{im})$ where $d_{ij} \subseteq D_j$.

The identity relation, defining the notion of redundancy in the ordinary database, is substituted in the fuzzy relational database with an explicitly declared similarity relation of which the identity relation is actually a special case.

A similarity relation, S(x, y), denoted also as xSy, for given domain D_j is a mapping of every pair of values in the particular domain onto the unit interval [0, 1], which reflects the level of similarity between them. A similarity relation is reflexive and symmetric as a traditional identity relation. However, a special form of transitivity is used:

$$\forall x,y,z \in D_j \quad S(x, z) \geq \mathrm{Max}\{\mathrm{Min}\,[S(x, y), S(y, z)]\}$$

Each of the attributes in the fuzzy database has its own similarity table, which includes the levels of similarity between all values appropriate for the particular attribute.

Rough Sets Background

Rough Set (RS) theory was introduced by Pawlak [6,7] as a mathematical tool for the analysis of *inconsistent or ambiguous* description of objects. The rough set philosophy is based on the assumption that every object in the universe *U* is associated with certain amount of information (data, knowledge). This information can be expressed by means of attributes describing the object. Objects, which have the same description are said to be indiscernible with respect to the available attributes.

The *indiscernibility relation* constitutes the mathematical basis of rough set theory. It induces a partition of the object domain into blocks of indiscernible objects, called elementary sets, which can be used to build knowledge about real or abstract worlds. Any subset *X* of the universe *U* may be expressed in terms of blocks either precisely or approximately. In the latter case, the subset *X* may be characterized by two ordinary sets, called the *lower* and *upper approximations*. A rough set is defined by means of these two approximations, which coincide in the case of an ordinary set.

The lower approximation of *X* is composed of all the elementary sets whose elements certainly belong to *X*, while the upper approximation of *X* consists of all the elementary sets whose elements may belong to *X*. The difference between the upper and lower approximation constitutes the boundary region of the rough set, whose elements cannot be characterized with certainty as belonging or not to *X* using the available information. The information about objects from the boundary region is, therefore, inconsistent or ambiguous.

In the domain of spatial data handling Schneider [8] and Worboys [9] used the RS theory to account for imprecision resulting from spatial or semantic data resolution. The work by Ahlqvist et al. [10] presented RS theory-based measures of uncertainty for spatial data classification.

In this paper we adopt the rough set theory for a problem of multiple criteria classification where class membership is induced by both the spatial relationship of containment and attribute relationship of indiscernibility.

Fuzzy Categorization

Assume we have a database of information about a specific application for which we are interested in developing a decision support system. Because of the nature of the data represented we utilize a fuzzy database approach based on the use of similarity relationship for the data domains of the N database attributes: A_1, A_2, A_3, … A_N which we will denote as a, b, c, …, p for notational simplicity in the following. If the data that is driving the system comes from a number of sources, then it possible that the same data will occur in these sources and so we will allow for the purposes of this application redundant (duplicate) tuples in the database.

Now to use this data it will be necessary to have the data entries (tuples) classified into the categories, C_1, C_2,…, C_T used in the decision support system. We will assume we have some access to human expertise and some small subset of the data has been classified by the expert for each of the respective categories. We will now address how to utilize these examples to drive the classification of the remaining data.

For each category C_i we have a corresponding set of tuples that have been classified as belonging to that category:

$$C_i{}^* = \{ t_1{}^i, t_2{}^i, .. t_m{}^i \}$$

where the tuple $t_j{}^i$ will be written as

$$t_j{}^i = (a_j{}^i, b_j{}^i, \ldots p_j{}^i)$$

In order to classify any remaining database tuple, t_k, we will proceed by considering the degree of matching of t_k with the tuples for each category allowing us to then decide into which category to classify the tuple.

Let us illustrate this in the case of a crisp database. Consider a category C for which the expert has classified 3 tuples:

$$\{ t_1 = (a_m, b_k, c_m, d_k), t_2 = (a_k, b_k, c_k, d_k), t_3 = (a_k, b_m, c_m, d_m)\}$$

The tuple to be classified is

$$t_k = (a_k, b_k, c_k, d_k)$$

By simple equality or inequality of the crisp data values we obtain a match count M

$$M(t_k, t_1) = 2, M(t_k, t_2) = 4, M(t_k, t_3) = 1$$

Normalizing by the number of elements in the tuple we have

$$M'(t_k, t_1) = 1/2, M'(t_k, t_2) = 1, M'(t_k, t_3) = 3/4$$

Since the tuple t_k here is an exact match to t_2, it would have clearly been classified by the expert as belonging to the category C. So in general we can use the maximum of the matching scores to represent the degree to which a tuple belongs to a particular category.

Now for our fuzzy database we must consider not the equality but the similarity of the attribute values in the tuples being matched. This will produce a similarity matching count M_{sim}. For example considering $M_{sim}(t_k, t_1)$, if $S(a_k, a_m) = 0.80$ and $S(c_k, c_m) = 0.65$ then we have these values instead of a zero for non-exactly matching attribute values. We can use a simple averaging of the similarity values to obtain a final similarity matching score. So

$$M'_{sim}(t_k, t_1) = (2 + 0.80 + 0.65)/ 4 = 3.45/4 = 0.86$$

We will now use this to approach our classification problem as one in which we are attempting to determine the membership degree μ of an unclassified tuple t in a particular category C. As discussed we can use the maximum of the similarity matches to all of the tuples that were classified by the expert as belonging to the category, that is, chose the closest match. Obviously if there is an exact similarity match, μ= 1 then the tuple t must be classified as belonging to the category C.

In general then for category C_i containing m tuples and considering tuple t_k

$$\mu_{C_i}(t_k) = \operatorname*{Max}_{l=1}^{m} [(S(a_k, a_j^l) + S(b_k, b_j^l) + \dots S(p_k, p_j^l))] / N$$

So if the maximum membership μ is for a category C_q, then we will classify t_k as belonging to C_q.

If we are only interested in crisp categories then we have a final answer. However for other applications it may be important to explicitly use the exact degree of membership as developed. This approach could also be combined with a relaxation of the classifications provided by experts permitting them to provide a degree of membership with each tuple they classify for each category. This may allow a more realistic characterization of the categories.

Rough Spatial Categorization

Consider a geographic space comprised of a set of points and a set of polygons such that each polygon contains a subset of points. Each point can be characterized by some attributes and each attribute has a defined value domain. Subsets of points indiscernible in terms of attribute values correspond to elementary sets in the sense of rough set theory.

We can partition the geographic space into polygons such that polygons characterized by the same property constitute a class. As an example consider a wildlife preserve partitioned into habitat areas based on the richness of species. Habitats characterized by the same richness of species constitute a class. Assume a survey of wildlife in the preserve revealed a distribution of species along with their attribute characteristics. We are then interested in learning whether the elementary sets of surveyed species describe the habitat classes precisely or approximately. Note that even though the description of habitat classes by surveyed species is based on attributes, it is facilitated by the containment relationship resulting from point-in-polygon intersection. The lower approximation of a habitat class is composed of all elementary sets that are fully contained in the class, while the upper approximation consists of all elementary sets, which are partially contained in the class. The partial containment means that only some members of the elementary set are contained in a given polygon class while other representatives are contained in other polygon classes.

Information about point-in-polygon pattern can be stored in a point attribute table called here the *classification table* where one column represents a polygon class and the rest of columns represent point attributes.

More formally, information about point-in-polygon pattern can be represented by a pair **A** = *(U, A)*, where *U* is a non-empty finite set of points obtained from point-in-polygon intersection, and $A = \{a_1, \ldots, a_n)$ is a non-empty finite set of *attributes*, i.e., $a_i : U \rightarrow V_a$ for $a \in A$, where V_a is called *value set* of the attribute a_i. In the classification table the attributes that belong to **A** are called *conditional attributes* or *conditions* and are assumed to be finite. The classification table is then a pair **A** = $(U, A\ U\{c\})$, where *c* represents a distinguished attribute called a *class*. The *i*-th class is a set of objects $C_i = \{o \in U: c(o) = c_i\}$, where c_i is the *i*-th class value taken from class value set $V_c = \{\ c_i, \ldots, c_{|V_c|}\ \}$.

This is almost identical to the fuzzy classification technique described above, however, here we are simply using the fuzzy equivalent of two alpha cut sets.

We can extend this technique through the establishment of a dominance relation based on preferences. In the classical rough set theory the data about objects belonging to a set *U* can either be quantitative or qualitative. The classical rough set approach is not able, however, to deal with preference-ordered attribute domains and preference-ordered classes. We can examine situations where domains of some attributes have an established preference order. Referring to the wildlife example, habitats can be ordered from poor, through satisfactory, to good, and species can be described by preference-ordered attributes such as abundance and fitness.

One technique typically used for achieving a similar, even automatic weighting technique is Ordered Weighted Averaging (OWA), described in [11]. This can help set up dominance relations, to achieve more even accurate classifications.

Acknowledgments

The authors would like to thank the Naval Research Laboratory's Base Program, Program Element No. 0602435N for sponsoring this research.

References

[1] Schmitz, A., Morris, A.: Modeling and Manipulating Fuzzy Regions: Strategies to Define the Topological Relation between Two Fuzzy Regions. Control and Cybernetics 35 #1, 73–95 (2006)

[2] Morris, A.: A Framework for Modeling Uncertainty in Spatial Databases. Transactions in GIS 7# 1, 83–101 (2003)

[3] Buckles, B., Petry, F.: A fuzzy representation of data for relational databases. Fuzzy Sets and Systems 7(3), 213–226 (1982)

[4] Petry, F.: Fuzzy Databases: Principles and Applications. Kluwer Academic Publishers, Boston (1996)

[5] Zadeh, L.: Similarity relations and fuzzy orderings. Information Sciences 3(2), 177–200 (1970)

[6] Pawlak, Z.: Rough sets. International Journal of Computer Information Sciences 11(5), 341–356 (1982)

[7] Pawlak, Z.: Rough sets - theoretical aspects of reasoning about data. Kluwer Academic Publishers, Dordecht (1991)

[8] Schneider, M.: Spatial Data Types for Database Systems. In: Schneider, M. (ed.) Spatial Data Types for Database Systems. LNCS, vol. 1288. Springer, Heidelberg (1997)

[9] Worboys, M.: Computation with imprecise geospatial data. Computers, Environment and Urban Systems 22, 85–106 (1998)

[10] Ahlqvist, O., Keukelaar, K., Oukbir, K.: Rough and fuzzy geographical data integration. International Journal of Geographical Information Science 17(3), 223–234 (2003)

[11] Yager, R.: Families of OWA operators. Fuzzy Sets and Systems 59, 125–148 (1993)

Supporting Spatial Decision Making by Means of Suitability Maps

Guy De Tré, Jozo Dujmović, and Nico Van de Weghe

Abstract. Spatial Decision Support Systems (SDSS) are interactive, computer-based systems, designed to support decision makers in achieving a higher effectiveness of decision making while solving a semi-structured spatial decision problem. Current spatial decision support techniques are predominantly based on boolean logic, which makes their expressive power inadequate. In this chapter it is presented how the Logic Scoring of Preference (LSP) method, helps to overcome the inadequacies present in traditional approaches. LSP is well suited to produce so-called dynamic, geographic *suitability maps* (S-maps), which provide specialised information on the suitability degree of selected geographic regions for a specific purpose. The presented approach is based on soft computing and many-valued logic.

1 Introduction

Advances in *geographical information systems* (GIS) have created very efficient techniques to collect, integrate and manage large amounts of geographical data in different forms and scales. Among the various GIS application fields one can consider the following three main categories [18]:

- *Socio-economic applications* like, e.g., urban or regional planning, planning for industrial development, agricultural land use, housing, education, recreation, land registry, archaeology, natural resources, etc.

Guy De Tré
Dept. of Telecommunications and Information Processing, Ghent University, St.-Pietersnieuwstraat 41, B-9000 Ghent, Belgium
e-mail: Guy.DeTre@UGent.be

Jozo Dujmović
Dept. of Computer Science, San Francisco State University, 1600 Holloway Ave, San Francisco, CA 94132, U.S.A.
e-mail: jozo@sfsu.edu

Nico Van de Weghe
Dept. of Geography, Ghent University, Krijgslaan 281 (S8), B-9000 Ghent, Belgium
e-mail: Nico.VandeWeghe@UGent.be

J. Kacprzyk et al. (Eds.): Uncer. Appro. for Spatial Data Model. and Process., SCI 271, pp. 9–27.
springerlink.com

- *Environmental applications* like, e.g., forestry, fire and epidemic control, flood, earthquake, hurricane and tsunami prediction, pollution control, etc.
- *Management applications* like, e.g., organization of pipeline networks and other services such as electricity and telecommunications, real-time vehicle navigation, planning of public services like health care, fire protection and security, etc.

In general GIS offer appropriate techniques for data management, information extraction, routine manipulation and visualisation, but they lack necessary analytical capabilities to efficiently support management and decision-making processes [10, 16]. For such purposes, a *spatial decision support systems* (SDSS) is used. A SDSS is generally defined as 'an interactive, computer-based system designed to support a user or group of users in achieving a higher effectiveness of decision making while solving a semi-structured spatial problem' [25]. As such, SDSSs have to provide insight into judgements of trade offs between various decision options. Because of the critical importance of visualisation, a SDSS is usually integrated with a GIS.

An important research area in SDSSs is that of spatial multiple criteria decision making (SMCDM) [19], also known as spatial multi-criteria evaluation [15]. The basic idea behind SMCDM is to evaluate multiple geographic choice alternatives using multiple decision criteria. Traditional SMCDM approaches are based on boolean logic (see, e.g., [14, 4, 9, 12, 17, 26, 29, 15]).

Ashley Morris was among the first to recognise that such traditional SMCDM approaches suffer from an inappropriate logical foundation, and an inadequate level of intelligence [20, 22]. The decision weighting techniques assign weights to the criteria somewhat arbitrarily and the criteria are strictly boolean. As a solution, he proposed to introduce more flexibility by using fuzzy logic and fuzzy set theory [30]. More specifically, he proposed analysis techniques that allow continuous or fuzzy functions to be assigned fuzzy values. Ashley Morris also recognised the importance and need for dynamically generated maps, which are able to instantly reflect the impact of a range of different parameter values and options on the decision making process.

The work presented in this chapter is motivated by the same observations: There is a need for advanced SMCDM techniques that are based on fuzzy logic and support the construction of specialised dynamically generated maps, which offer decision makers information about the overall suitability of a specific area for a selected type of use. Typical examples of such use are

- construction of industrial objects, homes, hospitals, schools, railway stations, airports, entertainment and sport centres;
- land use planning and management of natural resources;
- forestry, fire and epidemic control, disaster prediction, pollution control;
- navigation support for vehicles, boats, planes and spacecrafts.

In all cases decision makers are interested to evaluate and compare locations or regions from the standpoint of their suitability.

Fuzzy logic has been successfully applied in spatial data analysis and multicriteria evaluation. Consider, e.g., [3, 20, 15, 24, 21, 27, 23, 1, 2, 28]. In this chapter we present a novel SMCDM technique. The presented work is an extension of what we presented in [8]. Central in the novel SMCDM technique is the concept of a *suitability map* (S-map). An S-map is defined as a spatial distribution of the overall degree of suitability for a specific use [8]. The overall degrees of suitability are computed by the soft computing Logic Scoring of Preference (LSP) method which allows to adequately reflect the flexible suitability criteria and knowledge of the decision makers. In fact, in a general case the degree of suitability depends on a variety of logic conditions that decision makers specify using reasoning techniques that are typical for soft computing. Each overall degree of suitability is represented by a real number of the unit interval $[0, 1]$, where the value 0 denotes an unsuitable location and the value 1 (or 100%) denotes the maximum level of suitability.

The overall suitability of a specific area for a selected type of use typically depends on a (finite) number of attributes, which are obtained or derived from traditional geographic maps that are managed by a GIS. Such maps represent the distribution of selected geometric objects (borderlines, cities, roads, railroads, airports, rivers, forests, etc.) in the two-dimensional space and other information that may be of interest for complex planning and decision making. Attributes may characterise physical characteristics of terrain (slope, altitude, material, distance from major roads, distance from green areas, distance from lakes, etc.), available infrastructure (supply of water, supply of electrical energy, sewage system, telecommunications, transport systems, etc.), urban characteristics (distance from major schools, shopping areas, entertainment, sport facilities, hospitals, the density of population, etc.), legal status (private property, governmental property, areas reserved for special activities), economic development (local industries, businesses, employment abilities), pollution (air, water, noise), etc.

For each relevant attribute, an adequate criterion is constructed by using soft computing techniques. The use of soft computing techniques guarantees the efficient representation and handling of the decision maker's domain knowledge about the criterion under consideration. Next, the criteria are evaluated for each geographic choice alternative. As such, a set of elementary suitability degrees is obtained for each alternative. These elementary suitability degrees are then aggregated using a flexible soft computing aggregation technique which is based on the use of the generalized conjunction/disjunction function [6, 7]. This technique guarantees the efficient representation and handling of the decision maker's domain knowledge. As such the relative importance of criteria and combination of criteria evaluations, which are necessary to determine the overall suitability degree of each cell or area under consideration, are modelled in a human consistent way.

Initial experiments have shown that the presented approach allows for a dynamic generation of suitability maps. This provides decision makes with flexible tools to analyse the impact of changes of parameter settings in 'realtime', hereby accurately reflecting their domain knowledge and experience in an 'intelligent' way. Such a process would otherwise take a lot of resources and is even almost impossible in case of manual data processing. This also illustrates that there is a clear need for SDSSs to provide, in a semantically rich way, the information necessary for advanced public and professional decision making related to urban planning, industrial development, corporate planning, etc. In particular, there is a need for soft computing suitability maps that show suitability indicators based on flexible suitability criteria that include sophisticated logic conditions.

The remainder of the chapter is organised as follows. In the next Section 2 the concept of an S-map is described. In Section 3 it is presented how S-maps are constructed. Subsequently, the creation of the attribute tree, the definition of elementary criteria, the creation of the aggregation structure and the computation of the overall suitability degree are dealt with. An illustrative example on terrain suitability for home construction is presented in Section 4. Next, GIS integration and some implementation issues are discussed in Section 5. Finally, in Section 6, a summary of the work and some general observations are presented.

2 Description of Suitability Maps

To reflect the overall suitability of a specific area for a selected type of use, the concept of an S-map is introduced. Underlying to an *S-map* is a field-based, raster model aimed to represent overall suitability degrees that are associated with given geographical locations. As illustrated in Figure 1, it is defined as a finite partition of a subspace of the two-dimensional space. This subspace is determined by the perpendicular horizontal X-axis and vertical Y-axis and covers the area under investigation. The elements of the partition are convex subsets, which are called *cells* and are the smallest geographical units that can be considered in the raster. In this paper, like in most geographical information systems, all cells are defined to have a square shape of length l. The coordinates (x,y) of a cell denote the centre of the cell.

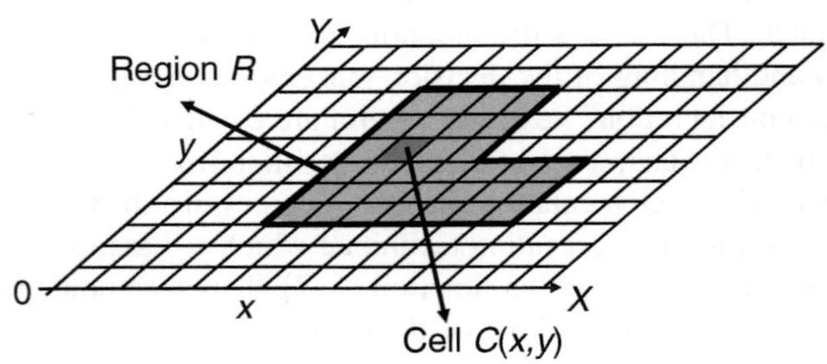

Fig. 1. Underlying raster model of S-maps.

This raster model is used to represent a spatial distribution of the overall degree of suitability of some analysed cells for a specific use. This is illustrated in Figure 2. Each overall degree of suitability is represented by a real number of the unit interval $[0,1]$ along a third Z-axis which is perpendicular to the (X,Y) space. Hereby, the value 0 denotes an unsuitable location and the value 1 (or 100%) denotes the maximum level of suitability. Alternative representations are possible. For example, a continuous grayscale can be used to represent the values of $[0,1]$. The value 0 then

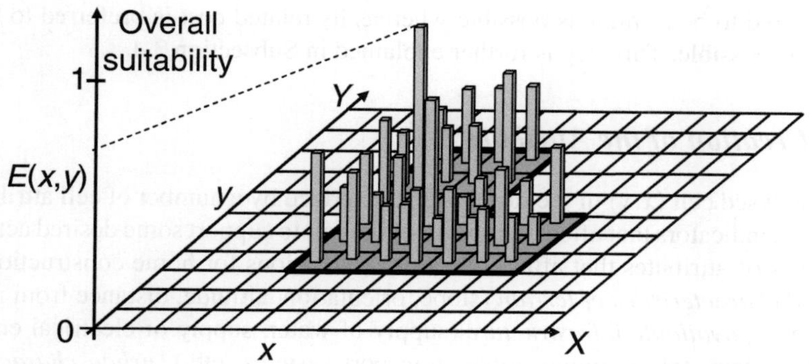

Fig. 2. An example of an S-map.

corresponds to the color white, while 1 corresponds to black. The closer the cell color is to black, the better overall suited is the cell.

3 Design of Suitability Maps

The design of an S-map is done, in accordance to the LSP method [6, 7], in subsequent steps which are handled in the next subsections. The main steps are:

1. *Creation of the attribute tree.* This tree contains and structures all parameters that affect the overall suitability. It is build by the decision maker. Attribute trees are described in Subsection 3.1
2. *Definition of elementary criteria.* The decision maker has to provide an elementary criterion for each attribute involved in the decision process. These criteria will be evaluated during S-map construction. For each analysed cell in the underlying raster model, the evaluation of each criterion will result in an elementary satisfaction degree. The definition of the elementary criteria is dealt with in Subsection 3.2.
3. *Creation of the aggregation structure.* For each analysed cell, all associated elementary satisfaction degrees must be aggregated. Therefore, the decision maker has to create an aggregation structure, which adequately reflects his domain knowledge and reasoning. This creation process is described in Subsection 3.3.
4. *Computation of the overall suitability degree.* Once the attribute tree, the elementary criteria and the aggregation structure are available, the S-map construction can start. The elementary criteria can be evaluated and their resulting elementary satisfaction degrees can be aggregated in order to compute the overall satisfaction degree of each analysed cell. In case of regions, the overall satisfaction degree of the region must be computed. If applicable, also financial, cost aspects will be taken into account at this stage. The overall suitability degree is finally computed taking into account that the overall satisfaction degree of a cell or region is

preferred to be as high as possible whereas its related cost is preferred to be as low as possible. This step is further explained in Subsection 3.4.

3.1 Creation of the Attribute Tree

Each analysed cell (x,y) of the raster is characterized by a number of cell attributes, which are indicators that affect the ability of the cell to support some desired activity. Examples of attributes that affect selection of locations for home construction are *physical characteristics of terrain* (slope, orientation, altitude, distance from major roads, etc.), *available infrastructure* (supply of water, supply of electrical energy, sewage system, telecommunications, transport systems, etc.), *urban characteristics* (distance from major schools, shopping areas, entertainment, sport facilities, hospitals, the density of population, etc.), *legal status* (private property, governmental property, areas reserved for special activities), *economic development* (local industries, businesses, employment opportunities), *pollution* (air, water, noise, odour), etc.

It's the task of the decision maker to select an appropriate, non-redundant, consistent array of n cell attributes $(a_1(x,y), a_2(x,y), \ldots, a_n(x,y))$, which are either mandatory or desired requirements in the decision making process. Each attribute is assumed to be a function of the coordinates (x,y). For simplicity this array will be denoted by $(a_1, a_2, \ldots, a_n)$.

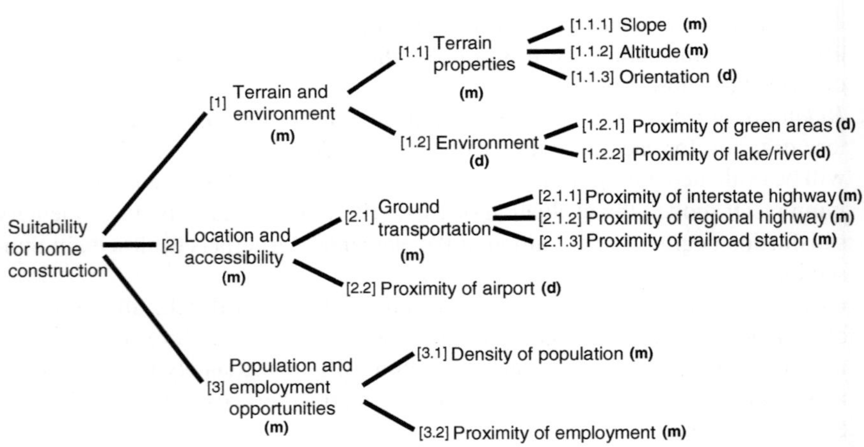

Fig. 3. An example of an attribute tree.

The selected attributes are organised in an attribute tree, in order to better reflect the decision maker's knowledge. As such criteria, can be subdivided in sub criteria. The selected cell attributes are then represented in the leaf nodes of the tree. Each sub criteria is labeled with a symbol '(m)' or '(d)' to denote whether the sub criteria is mandatory or desired. An example of an attribute tree for terrain selection for home construction is given in Figure 3. As a shorthand notation, each node is assigned a unique number code, which precedes the node name in the figure.

3.2 Definition of Elementary Criteria

For each selected cell attribute in $(a_1, a_2, \ldots, a_n)$, an elementary attribute criterion must be defined. This criterion should represent the requirements for the values of the attribute with respect to the specific use of the decision support model under construction. Soft computing techniques are used to adequately reflect the decision maker's expertise in this field.

More specifically, each elementary criterion is modelled by means of a *fuzzy set* [30] that is defined over the set of valid values for the attribute, i.e., the domain of the attribute. The membership function

$$\mu_{a_i} : dom_{a_i} \rightarrow [0,1]$$

of the criterion for attribute a_i, $i = 1, 2, \ldots, n$ then specifies the level of satisfaction of each potential value of the attribute. A membership grade $\mu_{a_i}(v_i) = 0$ denotes that the value v_i is fully unsatisfactory, a membership degree $\mu_{a_i}(v_i) = 1$ denotes that the value v_i is fully satisfactory, whereas each other value $\mu_{a_i}(v_i) \in]0,1[$ denotes a partial satisfaction of v_i. As such, membership grades are interpreted as degrees of compatibility [5], i.e., a membership grade $\mu_{a_i}(v_i)$ expresses to which extent the domain value a_i is compatible with the 'prototype'-elements for a_i that are required/desired by the decision maker.

An example of elementary attribute criteria for the attributes of the attribute tree of Figure 3 is shown in Figure 4. With these criteria it is for example specified that the ideal distance from the home construction location to a regional highway is considered to be from 100 to 200 meters. If the distance is greater than 2000 meters or less than 25 meters this is considered unacceptable.

3.3 Creation of the Aggregation Structure

To compute the overall suitability degree of a given cell, all elementary attribute criteria have to be evaluated using the actual attribute values of the cell. To evaluate the criterion for attribute a_i, $i = 1, 2, \ldots, n$ the membership grade $\mu_{a_i}(v_i)$ of the actual value v_i of a_i is determined. In case of piecewise linear membership functions, as in Figure 4, criterion evaluation can be done straightforwardly by linear interpolation.

The value $\mu_{a_i}(v_i)$ reflects the elementary degree of satisfaction of the cell, which is obtained if only the criterion for attribute a_i is considered. As such, the evaluation of the n criteria for the attributes $(a_1, a_2, \ldots, a_n)$ results in an array

$$(\mu_{a_1}(v_1), \mu_{a_2}(v_2), \ldots, \mu_{a_n}(v_n)) \in [0,1]^n$$

of n elementary satisfaction degrees. Next, these n elementary satisfaction degrees must be aggregated to generate an overall satisfaction degree of the cell.

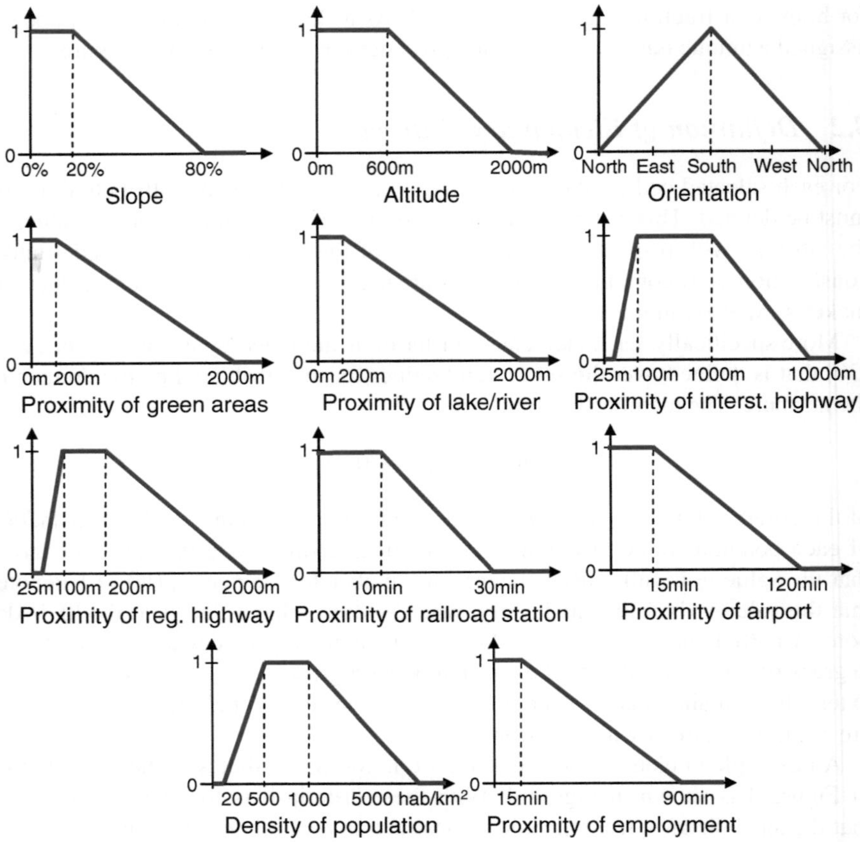

Fig. 4. An example of elementary attribute criteria.

Because fuzzy set theory is used in the criteria evaluation process, the aggregation must be based on a continuous logical foundation (else information loss occurs). Furthermore, it is very important that the aggregation reflects the decision maker's reasoning as accurate as possible. Therefore, a solution where the decision maker interactively builds up the *aggregation structure* is aimed for. The LSP method [6, 7] offers such facilities. This is why the presented approach is based on LSP.

In Subsection 3.3.1, the simple LSP aggregators are described. Simple LSP aggregators can be used to compose more complex aggregators. An example of such a compound aggregator, the conjunctive partial absorption operator, is presented in Subsection 3.3.2. Simple and compound aggregators are the basic building blocks of the aggregation structure. The aggregation structure should reflect the structure of the attribute tree. In Subsection 3.3.3 the construction of the aggregation structure is discussed.

3.3.1 Simple LSP Aggregators

The simple LSP aggregators are all graded preference logic functions and based on a superposition of the fundamental *Generalized Conjunction/Disjunction* (GCD) function [7]. The parameter r of the GCD function determines its logical behavior. As such a continuous variety of logical functions ranging from full conjunction to full disjunction can be modelled.

Furthermore, the GCD function is a weighted mean. This implies that each of its arguments x_i, $i = 1,2,\ldots,n$ is parameterized with an associated weight w_i. These weights denote the relative importance of the argument, i.e., the relative importance of the corresponding attribute criterion, within the decision support process. The semantics of the weights are defined as follows:

- Each weight is represented by a real number between 0 and 1, i.e., $0 < w_i < 1$, $i = 1,2,\ldots,n$. Weights 0 and 1 are impossible. Indeed, a weight 0 means that there is no input from this argument, so the argument must be excluded from the aggregation structure. A weight 1 implies that the aggregator has no other inputs, which makes the aggregator redundant and is neither allowed.
- Each edge in the attribute tree is assigned an associated weight. The weights of all edges of a given node must sum up to one.
- The larger the weight, the more important its associated argument is for the aggregation.

In LSP, the GCD function is implemented as a weighted power means (WPM). More specifically, the GCD function is implemented by

$$GCD : [0,1]^n \to [0,1]$$
$$(x_1, x_2, \ldots, x_n) \mapsto (w_1.x_1^r + w_2.x_2^r + \cdots + w_n.x_n^r)^{1/r}$$

where parameter $r \in [-\infty, +\infty]$ is the WPM exponent, $0 < w_i < 1$, $i = 1,2,\ldots,n$ are the associated weights (for which $\sum_{i=1}^{n} w_i = 1$) and n is the number of arguments.

For practical reasons, only 7 discrete levels of andness/orness are considered: conjunction, strong/weak partial conjunction, the arithmetic mean, strong/weak partial disjunction, and disjunction. This allows to precompute the parameter r and to

Operator	Symbol	Exponent r	
Full conjunction (**and**)	**C**	-∞	↑ simultaneity
Strong partial conjunction	**C+**	-3.510	
Weak partial conjunction	**C-**	0.261	
Arithmetic mean	**A**	1	neutral
Weak partial disjunction	**D-**	2.018	replaceability ↓
Strong partial disjunction	**D+**	9.521	
Full disjunction (**or**)	**D**	+∞	

Fig. 5. Seven discrete levels of andness/orness.

associate a linguistic label (and symbol) to each level. The 7 levels of andness/orness and their corresponding label, symbol and parameter value for r are given in Figure 5. Precomputation has been done by a software tool that computes r from a training set of desired input-output pairs. The decision maker then only has to choose among the 7 linguistic terms, which kind of andness/orness best corresponds to his or her needs. The ability to efficiently support such a discretization in a finite number of andness/orness levels is a main motivation to choose for the WPM implementation.

3.3.2 Example of a Compound LSP Aggregator

The simple LSP aggregators can be used to construct more complex, composed operators. To illustrate this, the conjunctive partial absorption (CPA) operators are presented in this subsection. These operators can be used to combine (or aggregate) satisfaction degrees that result from the evaluation of a mandatory and a desired criterion.

The required behavior of a CPA operator is presented in Figure 6. Herewith, it is reflected that the mandatory input (x) must be (at least partially) satisfied and that an insufficient satisfaction of the mandatory input ($0 < x < 1$) can be partially compensated by the desired input (y). As can be seen in the schema on top of the table in Figure 6, a CPA operator is composed of two simple LSP aggregators: a disjunctive operator ∇ and a conjunctive operator $\triangle$.

The implementation of a CPA operator is also based on a weighted power means (WPM). More specifically CPA operators are implemented by

$$\begin{aligned} CPA &: [0,1] \times [0,1] \to [0,1] \\ &(x,y) \mapsto ((1-w_x)((1-w_y).x^q + w_y.y^q)^{q/r} + w_x.x^r)^{1/r} \end{aligned}$$

where $q \in [-\infty, +\infty]$ is the andness/orness parameter of the disjunctive operator ∇, $r \in [-\infty, +\infty]$ is the andness/orness parameter of the conjunctive operator $\triangle$, and w_x and w_y are the weights that are associated with the inputs.

In case of a CPA operator, the decision maker has to provide the desired penalty and reward and has to select an appropriate elementary disjunctive operator ∇, and

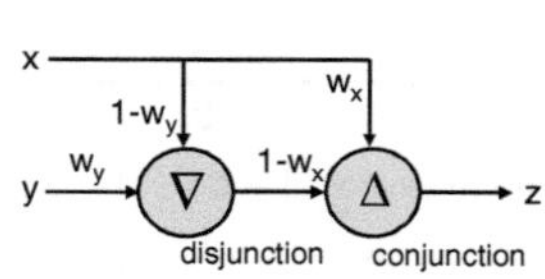

Mandatory input x	Desired input y	Output z
0	y (any value)	0
0 < x < 1	1 (perfect)	x + reward
0 < x ≤ 1	0 (worthless)	x - penalty

Fig. 6. Required behavior of a CPA operator.

conjunctive operator $\triangle$. Based on this input, the weights w_x and w_y and the andness/orness parameters p and q are precomputed as to obtain (or approximate) the required operator behavior. For the precomputation, a software tool that computes the parameters from a training set of desired input-output pairs is used.

3.3.3 Constructing the Aggregation Structure

The simple and compound LSP aggregators are the building blocks of the aggregation structure. The decision maker can use them to construct an easily understandable aggregation schema which is consistent with observable properties of human reasoning in the area of evaluation. This guarantees that the presented method has better facilities to model expert reasoning than traditional SMCDM approaches offer.

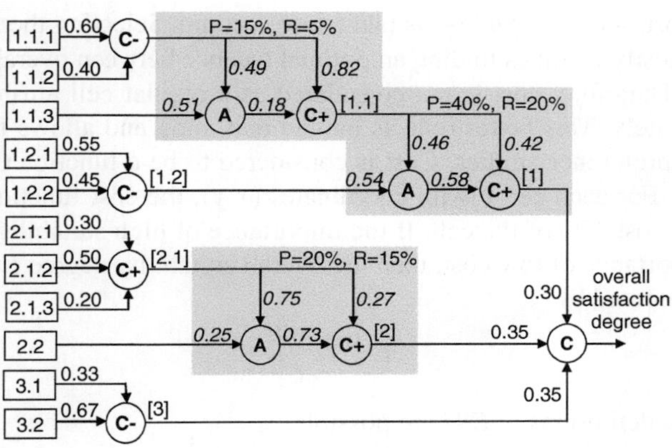

Fig. 7. Example of an aggregation structure.

In Figure 7 an example of an aggregation structure for selection of locations for home construction is given. This aggregation structure is conform with the attribute tree of Figure 3. The tree consists of five simple LSP aggregators and three CPA aggregators (denoted with shaded rectangles). For each CPA aggregator, a penalty P and reward R is given. The arrows in the structure have associated weights. Weights in *italic* font are precomputed, weights in regular font are provided by the decision maker. Some arrows are labeled with the unique node code of their corresponding node in the attribute tree. This visualises the correspondences between the attribute tree and the aggregation structure.

3.4 Computation of the Overall Suitability Degree

For S-map construction, the overall suitability degree of each analysed cell c with coordinates (x, y) must be computed.

1. Firstly, the n criteria for the n relevant attributes $(a_1, a_2, \ldots, a_n)$ must be evaluated, hereby using the n actual attribute values $(v_1^{(c)}, v_2^{(c)}, \ldots, v_n^{(c)})$ of the cell c. (For the sake of this explanation it is assumed that all these values are available and accurate. The evaluation results in an array

$$(\mu_{a_1}(v_1^{(c)}), \mu_{a_2}(v_2^{(c)}), \ldots, \mu_{a_n}(v_n^{(c)})) \in [0,1]^n$$

of n elementary satisfaction degrees.
2. Secondly, the elementary satisfaction degrees are evaluated using the aggregation structure of the decision problem. This results in the overall satisfaction degree

$$s^{(c)} \in [0,1]$$

of c.
3. Thirdly, cost —if applicable— is taken into account. Selecting the most suitable location mostly involves finding an optimal balance between overall satisfaction and cost. Therefore, cost is not considered as a regular cell attribute but dealt with separately. This better reflects human reasoning and allows for more efficient cost/preference studies. Cost is considered to be a function C of the analysed cells. For each cell c with coordinates (x,y), the cost function returns the associated cost $C^{(c)}$ of the cell. If the importance of high suitability is the same as the importance of low cost, then the overall suitability degree $E^{(c)}$ of the cell can be computed by

$$E^{(c)} = \frac{s^{(c)}}{C^{(c)}}.$$

Alternative definitions of $E^{(c)}$ are possible.

In case the overall suitability of a region R must be computed, the overall suitability degrees $E^{(c)}$ of all locations (cells) c of R must be aggregated. The averaging technique can be used for this purpose, in which case

$$E^{(R)} = \frac{\sum_{c\in R} E^{(c)}}{\sum_{c\in R} 1}.$$

where $\sum_{c\in R} 1$ equals the number of cells in R.

4 Illustrative Example

As an example consider the situation where five locations $L_1(x_1,y_1)$, $L_2(x_2,y_2)$, $L_3(x_3,y_3)$, $L_4(x_4,y_4)$ and $L_5(x_5,y_5)$ must be compared to each other with respect to their suitability for home construction. The selected attribute tree is the one that is presented in Figure 3. The elementary attribute criteria are defined as given in Figure 4 and the aggregation structure is the one presented in Figure 7. The actual attribute values, obtained from a GIS system, are presented in Table 1. Location L_5

Table 1. Input attribute values and costs for the five competitive locations

L	[1.1.1]	[1.1.2]	[1.1.3]	[1.2.1]	[1.2.2]	[2.1.1]	[2.1.2]	[2.1.3]	[2.2]	[3.1]	[3.2]	Cost (C)
L_1	40	700	SE	3000	1000	100	50	15	20	4000	5	2
L_2	5	500	S	5000	7000	5000	200	10	60	2000	15	1.7
L_3	18	1200	N	1000	2000	7000	120	7	15	500	30	1.2
L_4	35	300	SW	2500	500	1500	150	20	45	700	25	1.4
L_5	50	300	S	200	1000	4000	50	25	90	100	60	1

Table 2. Resulting overall satisfaction degrees $s^{(L)}$ and overall suitability degrees $E^{(L)}$

L	[1.1]	[1.2]	[2.1]	[1]	[2]	[3]	$s^{(L)}$	$E^{(L)} = \frac{s^{(L)}}{C}$
L_1	0.79	0.03	0.40	0.86	0.56	0.67	0.56	0.28 [45%]
L_2	1.00	0.00	0.73	0.96	0.87	0.91	0.87	0.51 [83%]
L_3	0.82	0.07	0.46	0.88	0.64	0.86	0.64	0.53 [86%]
L_4	0.86	0.05	0.72	0.90	0.87	0.91	0.87	0.62 [100%]
L_5	0.70	0.78	0.33	0.81	0.46	0.31	0.31	0.31 [50%]

is the cheapest, L_3 is 20% more expensive than L_5, L_4 is 40% more expensive than L_5, L_2 is 70% more expensive than L_5, and L_1 is 200% more expensive than L_5.

The resulting (intermediate) satisfaction degrees that obtained from the evaluation of the aggregation structure are presented in Table 2. The overall satisfaction degree $s^{(L)}$ of each location L is given in the second last column, whereas the overall suitability degree $E^{(L)}$ of each location L is given in the last column.

Based on their overall satisfaction degree, locations L_2 and L_4 are equally convenient. However, if cost is taken into account, then location L_4 turns out to be the best choice.

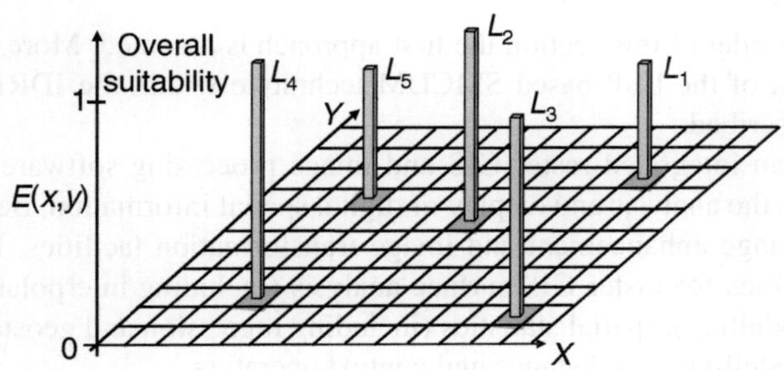

Fig. 8. S-map for selection of location for home construction.

The S-map of the example is presented in Figure 8. In real situations much more locations will usually be considered as potential options. Thanks to their integration with a GIS, S-maps allow to visualise the results of the decision making process in a compact way that is easy to interpret. Indeed, the overall suitability degrees can be integrated with other contextual, geographical information like borderlines, cities, roads, railroads, airports, rivers, forests, etc. The decision maker can be provided with a graphical user interface (GUI) in which he or she can select the contextual information to be presented in an interactive way.

Moreover, S-maps can be dynamically generated, which means that the decision maker is also provided with a GUI which allows to adapt the attribute criteria and the weight, penalty, and reward parameters of the aggregation structure. Once such a modification is done, the S-map can be reconstructed in order to instantly reflect the impact of the modification on the decision making process. This is a definite advantage of the presented approach: the user can dynamically modify the criteria or adjust values or perspectives; the presentation bars or colors can be calculated and be displayed dynamically and continuously.

In the next section, some preliminary test results and implementation issues are discussed.

5 GIS Integration

The integration of the SMCDM technique for S-map construction presented in the preceding sections and a GIS is crucial to use S-maps efficiently in practice. The GIS not only provides the necessary attribute data for the criteria evaluation, but also provides adequate visualisation facilities. Essentially, three integration approaches are possible [22]:

1. The first approach is to integrate the SMCDM technique within the GIS. This is for example the approach taken in [15].
2. The second approach is to implement GIS techniques and tools in SMCDM software. This is for example done in [11].
3. The third approach is to integrate both the SMCDM and GIS techniques at the operating system level [13].

In the remainder of this section the first approach is assumed. More specifically, the integration of the LSP based SMCDM technique within the IDRISI software package is described.

IDRISI is an integrated raster GIS and image processing software which provides tools for the analysis and display of digital spatial information. Besides image restoration, image enhancement and image transformation facilities, IDRISI also provides facilities for raster GIS surface analysis (including interpolation and hydrological modelling), spatial statistics (including regression and geostatistics), geographical modelling, and distance and context operators.

Integrating LSP facilities in IDRISI allows to use IDRISI's graphical and GIS facilities as efficient as possible. At the one hand this significantly reduces development time. At the other hand this makes the software solution tightly integrated and thus dependent on IDRISI, which is a commercial software package.

The rationale of the implementation of LSP facilities in IDRISI is sketched in Figure 9. IDRISI offers facilities to combine existing raster maps or to generate a new raster map from an existing map in an interactive way. Essentially the user has

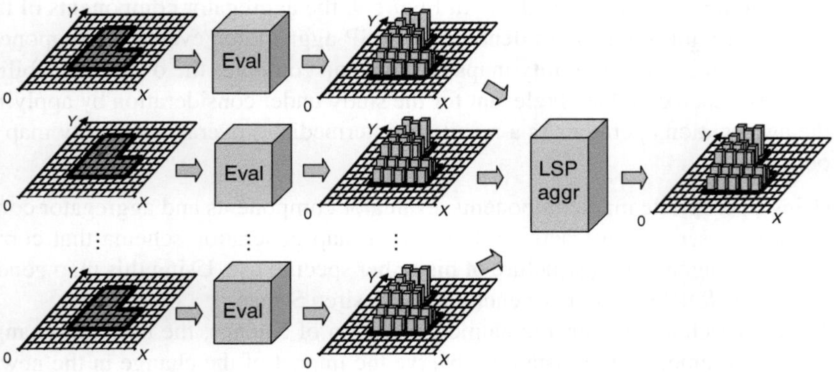

Fig. 9. LSP facilities in IDRISI.

to build up the map generation schema via drop and drag operations in a graphical user interface. The basic building blocks of a map generation schema are components that represent maps and components that represent operators. User defined operators are supported by IDRISI's COM/API interfaces. IDRISI's Application Programming Interface (API) allows for the development of stand-alone modules, written in C++ programming language, as add-ons to IDRISI. IDRISI can also be accessed via the industry-standard COM object model interface. Using COM, client applications can be written that control all aspects of IDRISI's operations.

For an initial prototype implementation, two specific types of generic operator components are implemented using C++ code: *evaluator components* and *aggregator components.*

- An *evaluator component* is used to generate an elementary suitability map from a regular raster map. Each evaluator component in a map generation schema corresponds to an evaluation function for an elementary criterion and is parameterized by the membership function μ_{a_i} of the criterion. In case of a trapezoidal membership function (as the ones depicted in Figure 4), only four parameters are required. In Figure 9, the evaluator components of the map generation schema are denoted by 'eval'. Each evaluator component takes as input a raster map that contains the relevant attribute data for the criterion under consideration. The component computes the elementary suitability degree of each cell that is relevant for the study under consideration by evaluating the criterion for the cell. As a result an elementary suitability map is obtained.
- An *aggregator component* combines a finite number (n) elementary suitability maps and generates a more general, aggregated suitability map. Each aggregator component in a map generation schema corresponds to a simple or compound LSP aggregator. In case of a simple aggregator, the component is parameterized by the precomputed andness/orness parameter r and a weight w_i, $i = 1, 2, \ldots, n$ for each of its inputs. A component that corresponds to a compound aggregator is parameterized by the precomputed andness/orness parameters r and q and

precomputed weights w_x and w_y. In Figure 9, the aggregator components of the map generation schema are denoted by 'LSP aggr'. Each evaluator component takes n elementary suitability maps as input and computes the overall suitability degree of each cell that is relevant for the study under consideration by applying the aggregation operator. As a result an (intermediate) overall suitability map is obtained.

Using appropriate map components, evaluator components and aggregator components, the user can interactively build up a map generation schema that corresponds to the aggregation structure of his or her specific use. Using this map generation schema IDRISI instantly generates the desired S-map.

The approach allows for a dynamic generation of S-maps: the user can change any of the parameters and instantly observe the impact of the change in the newly generated S-map.

To deal with criteria evaluation that is based on distance, IDRISI's distance operators can be directly used within the evaluator component. Moreover, IDRISI's interpolation facilities allow to deal with cases of missing or imprecise attribute data.

6 Conclusion

In this chapter it is presented how suitability maps (or S-maps) can be constructed and used to support spatial multicriteria decision making. An S-map is a specialised geographic map that represents a spatial distribution of the overall degree of suitability of selected cells (or areas) for a specific type of use.

S-maps are constructed using LSP methodology which implies that the main steps in S-map construction are:

1. Creation of an attribute tree.
2. Definition of elementary criteria.
3. Creation of an aggregation structure.
4. Computation of overall suitability degrees.

The elementary criteria and the aggregation structure are constructed using fuzzy set theory and soft computing techniques. This guarantees the efficient representation and handling of the decision maker's domain knowledge: relative importance of criteria and combination of criteria evaluations, which are necessary to determine the overall suitability degree of each cell or area under consideration, are dealt with in a human consistent way.

Advantages of S-maps can be summarised as follows:

- S-maps are general and flexible in the sense that they can express the suitability of the analysed geographic area for any specific use.
- The method of generating S-maps offers a high level of many-valued logic versatility originating from the LSP-based soft computing approach. It is easily understandable and consistent with observable properties of human reasoning in the area of evaluation.

- LSP models of suitability generate correct logic results in all points of the attribute space. The accuracy of such models cannot be reduced by unpredictable variations of attribute values. Therefore, the expected reliability of S-maps is very good.
- S-maps are dynamically generated from GIS databases.
- Users of S-maps can experiment with various suitability criteria and dynamically investigate effects of changing their parameters.

S-maps create various opportunities for future work. The initial efforts should be focused on improving the availability and reliability of input attribute data. There is also space for improving methods for working with incomplete and imprecise attributes of 'fuzzy' geographical databases.

With this chapter we contribute to the important research fields of fuzzy SMCDM and GIS. These fields also belonged to the research domain of Ashley Morris. We also indicated how the presented techniques could be implemented and integrated in a GIS framework. The potential applications of the presented technology are manifold. In future research, we aim to focus on applications that contribute to humanity, peace, and security. Three keywords that were also very important to Ashley and even characterise the driving forces behind Ashley's research activities. Ashley, we miss you but are convinced that you will always inspire us to never give up and reach our objectives.

References

1. Bone, C., Dragicevic, S., Roberts, A.: Integrating high resolution remote sensing, GIS and fuzzy set theory for identifying susceptibility areas of forest insect infestations. International Journal of Remote Sensing 26(21), 4809–4828 (2005)
2. Bordogna, G., Pagani, M., Pasi, G.: A flexible decision support approach to model ill-defined knowledge in GIS. In: Morris, A., Kokhan, S. (eds.) Geographical Uncertainty in Environmental Security, Dordrecht, The Netherlands. NATO Science for Peace and Security Series, pp. 133–152. Springer, Heidelberg (2007)
3. Burrough, P.A., Frank, A.U. (eds.): Geographic Objects with Indeterminate Boundaries. GISDATA series. Taylor & Francis, London (1996)
4. Carver, S.: Integrating multicriteria evaluation with GIS. International Journal of Geographical Information Science 5, 321–339 (1991)
5. Dubois, D., Prade, H.: The three semantics of fuzzy sets. Fuzzy Sets and Systems 90(2), 141–150 (1997)
6. Dujmović, J.J.: Preference Logic for System Evaluation. IEEE Transactions on Fuzzy Systems 15(6), 1082–1099 (2007)
7. Dujmović, J.J.: Characteristic Forms of Generalized Conjunction/Disjunction. In: Proceedings of the IEEE World Congress on Computational Intelligence, Hong Kong (2008)
8. Dujmović, J.J., De Tré, G., Van de Weghe, N.: Suitability Maps Based on the LSP Method. In: Torra, V., Narukawa, Y. (eds.) MDAI 2008. LNCS (LNAI), vol. 5285, pp. 15–25. Springer, Heidelberg (2008)
9. Faber, B., Wallace, W., Cuthbertson, J.: Advances in collaborative GIS for land resource negotiation. In: Proceedings of the 9th Annual Symposium on Geographic Information Systems (GIS 1995), Vancouver, Canada, pp. 183–189 (1995)

10. Fisher, M.M.: Expert systems and artificial neural networks for spatial analysis and modelling. Geographical Systems 1(1), 221–235 (1994)
11. Fisher, G., Markowski, M., Antoine, J.: Multiple criteria land use analysis, Working paper WP-96-006. International Institute for Applied Systems Analysis, Laxenburg, Austria (1996)
12. Jankowski, P.: Integrating geographic information systems and multiple criteria decision making methods. International Journal of Geographical Information Systems 9(3), 251–273 (1995)
13. Jankowski, P., Nyerges, T.L., Smith, A., Moore, T.J., Horvath, E.: Spatial Group Choice: a SDSS tool for collaborative spatial decision making. Geographical Systems 11(6), 577–602 (1997)
14. Jansen, R., Rietveld, P.: Multi-criteria analysis and geographical information systems: an application to agricultural land use in The Netherlands. In: Scholten, H.G., Stillwell, J.C.H. (eds.) Geographical Information Systems for Urban and Regional Planning, pp. 129–139. Kluwer Academic Publishers, Dordrecht (1990)
15. Jiang, H., Eastman, J.R.: Application of fuzzy measures in multi-criteria evaluation in GIS. International Journal of Geographical Information Science 14(2), 173–184 (2000)
16. Leung, Y., Leung, K.S.: An intelligent expert system shell for knowledge-based geographical information systems: 1. the tools, 2. some applications. International Journal of Geographical Information Systems 7(3), 189–213 (1993)
17. Lotov, A.V., Bushenov, V.A., Chernov, A.V., Gusev, D.V., Kamenev, G.K.: Internet GIS and Interactive Decision Maps. Journal of Geographical Information and Decision Analysis 1(2), 118–143 (1997)
18. Maguire, D.J., Goodchild, M.F., Rhind, D.W.: Geographical Information Systems: Principles and Applications. Longman Scientific and Technical (1991)
19. Malczewski, J.: GIS and multicriteria decision analysis. John Wiley & Sons, New York (1999)
20. Morris, A., Jankowski, P.: Combining fuzzy sets and databases in multiple criteria spatial decision making. In: Larsen, H.L., Kacprzyk, J., Zadrozny, S., Andreasen, T., Christiansen, H. (eds.) Flexible Query Answering Systems. Advances in Soft Computing, pp. 103–116. Physica-Verlag, Heidelberg (2000)
21. Morris, A., Jankowski, P.: Fuzzy techniques for multiple criteria decision making in GIS. In: Proceedings of the joint 9th IFSA World congress and 20th NAFIPS International conference (2001)
22. Morris, A., Jankowski, P.: Spatial decision making using fuzzy GIS. In: Petry, F.E., Robinson, V.B., Cobb, M. (eds.) Fuzzy modeling with spatial information for geographic problems, pp. 275–298. Springer, Heidelberg (2005)
23. Robinson, V.B.: A perspective on the fundamentals of fuzzy sets and their use in geographic information systems. Transactions in GIS 7(1), 3–30 (2003)
24. Scott, M.D., Robinson, V.B.: A multiple criteria decision support system for testing integrated environmental models. Fuzzy Sets and Systems 113, 53–67 (2000)
25. Sprague, R.H., Carlson, E.D.: Building effective Decision Support Systems. Prentice-Hall Inc., Englewood Cliffs (1982)
26. Tkah, J.R., Simonovic, P.S.: A new approach in to multi-criteria decision making in water resources. Journal of Geographical Information and Decision Analysis 1(1), 25–43 (1997)
27. Tran, L.T., Knight, C.G., O'Neill, R.V., Smith, E.R., Riitters, K.H., Wickham, J.: Environmental assessment, fuzzy decision analysis of integrated environmental vulnerability assessment of the Mid-Atlantic region. Environmental Monitoring 29(6), 845–859 (2002)

28. Wood, L.J., Dragicevic, S.: GIS-based multicriteria evaluation and fuzzy sets to identify priority sites for marine protection. Biodiversity Conservation 16, 2539–2558 (2007)
29. Wu, F.: SimLand: A prototype to simulate land conversion through the integrated GIS and CA with AHP-derived transition rules. International Journal of Geographical Information Science 12(1), 63–82 (1998)
30. Zadeh, L.A.: Fuzzy Sets. Information and Control 8(3), 338–353 (1965)

Exploring the Sensitivity of Fuzzy Decision Models to Landscape Information Inputs in a Spatially Explicit Individual-Based Ecological Model

Vincent B. Robinson

1 Introduction

This is part of an ongoing exploration of incorporating fuzzy logic into spatially explicit, individual-based ecological models of dispersal. Following the theoretical discussion of Robinson (2002), a prototypical model of small mammal dispersal behavior was used to demonstrate how the fuzzy control of dispersal agents could be implemented (Robinson and Graniero 2005a). The implementation showed how the Extensible Component Objects for Constructing Observable Simulation Models (ECO-COSM) system could be loosely coupled with geographic information system (GIS) database for spatially explicit ecological simulation modeling of individual behavior (Graniero and Robinson 2006). If the problem is viewed from a geocomputational management perspective, we can say that an animal agent must be able to query the state of relevant GIS layers within its local perceptual range and use that information to make decisions regarding its movement behavior. Its movement behavior inturn leads eventually to a change in the state of the agent. Within the ECO-COSM framework, this is handled by the Probe mechanism. By obtaining Probes from relevant Probeable landscape layers , an agent can acquire a perceptual inventory of its world (Graniero and Robinson 2006). Thus, the general approach is consistent with Bian's (2003) hybrid approach to representing the world in individual-based modeling, which incorporates a traditional grid model of the environment and an object-oriented model of individual organisms.

A fuzzy formulation of the dispersal model was used to implement four different fuzzy decision models. Robinson and Graniero (2005a) used a realistic landscape to illustrate similarities and differences in movement behavior as a function of crisp, compensatory, noncompensatory aggregation operators along with a

Vincent B. Robinson
Department of Geography
University of Toronto
Mississauga, ON L5L 1C6
Canada
e-mail: doc.robinson@utoronto.ca

J. Kacprzyk et al. (Eds.): Uncer. Appro. for Spatial Data Model. and Process., SCI 271, pp. 29–42.
springerlink.com

corresponding crisp equivalent. Simulations using the ECO-COSM model were used to compare fuzzy versus crisp model behaviors. However, like many fuzzy-based models there are number of important parameters that modulate the operation of the model. To-date there has been no investigation into how some of the more important parameters may, or may not, affect the behavior of the agents in the simulations. This paper presents a preliminary investigation into the sensitivity of results to variations in selected parameters that may most directly affect the information inputs to individual agents. In particular, the sensitivity of results to variations in the definition of the perceptual range is explored in relation to different decision models.

2 Information-Based Fuzzy Dispersal Model

An overview of the fuzzy dispersal model is presented here. Note that additional details on the model can be found in Robinson and Graniero (2005a,b) while Graniero and Robinson (2006) present more details on the implementation of Probes within the ECO-COSM framework. The dispersal movement behavior of an individual object is modeled as a function of a movement decision and a residence decision. When an object is to move from its current location it must decide on a target location. Once at the new, target, location it assesses its surroundings by gathering information used in the residence decision. In other words, has the object found a suitable location upon which to base a home range? If it finds the location not suitable, then it engages in movement decision making. Both movement and residence decisions were formulated as a fuzzy decision model where relevant goals and constraints are expressed in terms of fuzzy sets. A decision is determined through an aggregation of the fuzzy sets (Bellman and Zadeh 1970, Klir and Yuan 1995).

In the movement decision model constraints consist of those locations that are within the visible perceptual range and those spatially separated from conspecifics. The goal of an individual is to find a location as near the edge of the perceptual range as possible that is considered acceptable habitat and fits the set of constraints. An important social constraint is distance from conspecifics. Thus the goal set is a function of the spatial arrangement of habitat, and the dispersal imperative Of particular concern in this work is the definition of the "ideal" perceptual range (P) that in combination with a measure of visibility will determine the nature of the landscape information used in the decision process. In particular, note that β and θ control the shape of the membership curve (Table 1).

Once the individual has moved to a location, it decides whether or not it is suitable for stopping its dispersal movement. In the residence decision model (Table 2) the object is constrained by whether or not its current location is sufficiently spatially separated from conspecifics that a home range can be established, while the goal is to have habitat of sufficient area. A decision rule is applied that leads to the individual either taking up residence at the location or attempting a move to another location. The residence decision functions in these simulations act primarily as a stopping rule. It also provides the point at which the results of dispersal can be linked to a spatially explicit population model.

To implement differing decision models, agent classes of CompSquirrel, Non-compSquirrel, YagerSquirrel, and CrispSquirrel were created to correspond respectively to agents using decision models based on compensatory, noncompensatory, Yager, and crisp aggregation methods. In addition, a nonfuzzy, or crisp, set version of the decision models was constructed. In Tables 1 and 2, whenever there

Table 1. Movement Decision Sets (Graniero and Robinson 2006). This table is based on Robinson and Graniero (2005a) where the rationale for specific parameters and function forms is discussed in more detail.

Equation	Description
$C^M = \Psi \cap F$	**Constraint Set(** C^M **)** constraining the search to those locations that are in the visible perceptual range(Ψ) and far from competing conspecifics (F).
$\Psi = P \cap L$	**Visible Perceptual Range (** Ψ **)** The degree to which a cell is both visible and falls within the perceptual range
$P = \mu_p(x) = \begin{cases} 1 & \text{if} \quad d_x^c \le \beta \\ \theta(\beta - d_x^c) + 1 & \text{if} \quad \beta < d_x^c < \beta + 1/\theta \\ 0 & \text{if} \quad \beta + 1/\theta \le d_x^c \end{cases}$	The fuzzy set defining the 'ideal' perceptual range for a single individual. $X = \{x\}$ is a finite set of locations bounded by the limits of the study area. d_x^c is the Euclidean distance from the location of the dispersing animal object, *c*, to location *x*. The point at which $\mu_p = 1$ is represented by β and the parameter θ controls the rate at which $\mu_p \to 0$.
$L = \mu_L(x) = \max(\min\left(\frac{los_x^c - \alpha}{\beta - \alpha}, \frac{\gamma - los_x^c}{\gamma - \beta}\right), 0)$	The fuzzy set describing the degree to which location *x* is visible to an individual. The membership function for L is defined by a closed-form triangular function where los_x^c is the angle at which location *x* is visible from location *c*. If the local terrain creates a physical obstruction to visibility between *c* and *x*, then $L = 0$.
$F(x) = \mu_F(x) = 1.0 - \left(\bigcup_{k=1}^{c} \mu_{NC}^k(x)\right)$	The fuzzy membership of each location in the set of **far_from_conspecific** where if a conspecific is within the visible perceptual range (*i.e.*, $k \in^{0+} \Psi$.) then d_i^k is the distance from conspecific *k* to location *i*.
$NC^k(x; \alpha, \beta) = \mu_{NC}^k(x) = \begin{cases} \frac{\beta - d_x^k}{\beta - \alpha} & \alpha \le d_x^k \le \beta \\ 0 & \text{otherwise} \end{cases}$	The fuzzy set **near_conspecific** *k* where $\mu_{NC}^k(x)$ is the degree to which *x* is near conspecific *k* and d_x^k is the distance from conspecific *k* to *x*.

Table 1. *(continued)*

Equation	Description
$G^M = A \cap \mathrm{I}$	**Goal Set(** G^M **)** degree to which a location is as near the edge of the perceptual range as possible and is forested.
$A(x) = \mu_A(x) = \begin{cases} 1 & if \quad forest \\ 0 & if \quad nonforest \end{cases}$	**Habitat**. In the case of this species that habitat would be forest. We use the crisp classification because it is unlikely, especially towards the edge of the perceptual range, that squirrels can evaluate vegetation in any detailed manner. Once an individual has moved to a location then, through exploratory movement, an evaluation of the habitat becomes more detailed.
$\mathrm{I}(x) = \mu_I(x) = \max\left(\min\left(1, \frac{d_x^c - \alpha}{\beta - \alpha}\right), 0\right)$	**Dispersal Imperative** membership function where $\alpha = 0$ and β is the distance of the farthest location in Ψ that has a non-zero membership value. Reflects the imperative of finding a home as far from the current location as possible., given constraint of perceptual range.
$D^M = C^M \cap G^M$	**Decision set** on first move, movement is to location with highest value. In case of ties, the first one in the list is chosen.
$B = \mu_B(x) = \left[\left(\left(\frac{\cos(q_p) + \cos(q(x))}{2}\right)^2 + \left(\frac{\sin(q_p) + \sin(q(x))}{2}\right)^2\right)^{0.5}\right]^{\rho}$	The fuzzy set representing the degree to which a location falls within the set of **direction_to_move**. where q_p is the direction, in radians, of the move to the current location and $q(x)$ the direction, in radians, from the current location (κ) to location x and exponent ρ functions like a *hedge*, we assume $\rho = 2$.
$D^M = (C^M \cap G^M) \cap B$	**Decision set** on subsequent moves. Movement is to location with highest value. In case of ties, the first one in the list is chosen.

is a connective, $\cup$ or $\cap$, then one of the aggregation methods is used. Robinson and Graniero (2005a) provide a detailed listing of the aggregation methods used at each level of the decision process by each class of agents. It is easily seen that each agent class would evaluate the landscape somewhat differently as a function

Table 2. Residence Decision Sets(Graniero and Robinson 2006). This table is based on Robinson and Graniero (2005a) where the rationale for specific parameters and function forms is discussed in more detail.

Equation	Description
$C^R = \bigcap_c \mu^c_{Far}(\kappa)$	The **Constraint Set(** C^R **)** is a function of the spatial separation from surrounding conspecifics.
$\mu^c_{Far}(\kappa) = \begin{cases} 1.0 - \left\{ \dfrac{1.0}{\left[1 + \dfrac{d_c(\kappa) - \beta^c_{Far}}{\theta^c_{Far} - \beta^c_{Far}} \right]} \right\} & if \;\; d_c(\kappa) \geq \beta^c_{Far} \\ 0.0 & if \;\; d_c(\kappa) < \beta^c_{Far} \end{cases}$	The membership of location κ in the fuzzy set **Far_from_conspecific** *c* where $d_c(\kappa)$ is the distance from conspecific c ($c = 1 \ldots k$) and the current location (κ) of the Agent, β^c_{Far} represents the limit of a hypothetical core and θ^c_{Far} is the distance at which membership = 0.5.
$G^R = LC \cap HA$	The degree to which location κ falls in the goal set G^R. In effect a measure of the degree to which the current animal location is habitat and contained within a large enough patch of habitat.
$LC(\kappa) = \mu_{LC}(\kappa) = \begin{cases} 1.0 & if \;\; oak \\ 0.9 & if \;\; oak\,/\,deciduous\_bottomland \\ 0.75 & if \;\; deciduous \\ 0.0 & if \;\; conifer \\ 0.0 & if \;\; early\_successional\_deciduous \\ 0.0 & if \;\; wetland, pasture, grassland, ag. \\ 0.0 & if \;\; water \end{cases}$	The degree to which a land cover type found in our GIS database can be considered quality habitat for a gray squirrel. The Agent uses the the land cover at the location, κ, where the squirrel has moved.
$HA(\kappa) = \mu_{HA}(\kappa) = \max\left(0, \min\left(1, \left[\dfrac{farea(\kappa) - \alpha_{HA}}{\beta_{HA} - \alpha_{HA}} \right]\right)\right)$	The degree to which location κ falls within the class of **minimum habitat area.** By setting the parameters $\alpha_{HA} = 0.3$ and $\beta_{HA} = 2.0$ any patch less than 0.3 ha is clearly too small while any patch greater than 2 ha is clearly large enough.
$D^R = G^R \cap C^R$	The membership of location κ in the **residence_location** set
IF $D^R \geq 0.5$ THEN **reside** ELSE **move**	The **decision rule** for residence versus move.

of the underlying decision model. For example, the compensatory method allows for lower memberships in one set to be compensated to some degree by higher memberships in the other which is not the case using the noncompensatory method.

3 Methodology

The approach to investigate the sensitivity of results to parameters controlling the perception of landscape by the agents includes:

1. The use of a landscape where 80% of potential locations are already occupied by a conspecific.
2. The use of the same study area and GIS database as Robinson and Graniero (2005a).
3. Varying the values of β and θ used to define the perceptual range of an agent.
4. Varying the maximum number of steps an agent can take before reaching a state of "dead".
5. Running simulations for the compensatory, noncompensatory, and Yager decision models.

All simulations will start with the same 84 agent locations.

3.1 Conspecific Landscape

One of the important spatially explicit variables in the simulation is the density of conspecifics. Territoriality can impede movement, especially if all suitable space is occupied and individuals are not able to cross undefended space. When the density of conspecifics reaches such a level, it may result in what is called a *social fence* (Wolff 1999). Robinson and Graniero (2005a) varied the distribution of conspecifics by creating a dense base distribution of conspecifics that occupied all potential home range locations. Then to generate different levels of density each variation was based on eliminating a certain percentage of conspecifics from the dense base distribution. In this study only the case where 20 percent were eliminated. Thus, providing a challenging environment where landscape information is crucial for agents to find their way to suitable locations.

3.2 The Landscape Database

This exploratory study uses the same landscape used in Robinson and Graniero (2005a,b). It will provide a base-line landscape upon which the results can be compared. The study area is an 11 km by 11km subset taken from a larger GIS database for Western Kentucky, USA. Species like the gray squirrel tend to have dispersal distances of less than 5,000 meters (Wolff 1999, Bowman et al. 2002). Hence, the study area is large enough to accommodate the simulation of gray squirrel natal dispersal movements.

Two data layers are used to model the habitat-relevant land cover and topography. The land cover layer is based on the Kentucky GAP Project and shows a gradation, moving west to east, from fragmented to nonfragmented oak/deciduous forest (Figure 1) . A digital elevation model (DEM) is used since topography is a determinant of how visible a location is to an animal object. Elevation data were generated from the USGS 7.5 minute Digital Elevation Models (DEM), with a cell size of 30m x 30m. Both the land cover and elevation data layers were made available to us by the Mid-America Remote Sensing Center (MARC).

3.3 Perceptual Range and Number of Steps

The perceptual range of individual agents determines the quantity, and quality, of the landscape information that is made available to the decision process. Perceptual range is the distance from which a particular landscape element can be perceived as such. The perceptual range represents the informational window onto the larger landscape. $P(x)$ determines the extent over which there is some information about the landscape that can be perceived and information is retrieved from the GIS database. Subsequent operations are confined to the area where $P(x) > 0$. The parameter β determines the distance at which the membership in $P(x)$ begins to decline. From the individual out to β the membership is 1.0. Since the resolution of the raster layers is 30m x 30m the values of β that will be used in these simulations are 30, 60, and 90. Holding θ constant, the area around the individual where $P(x) > 0$ increases as β increases (Figure 2). However, since it is a membership curve it also affects the degree to which other aspects of the landscape are presumed to be perceived.

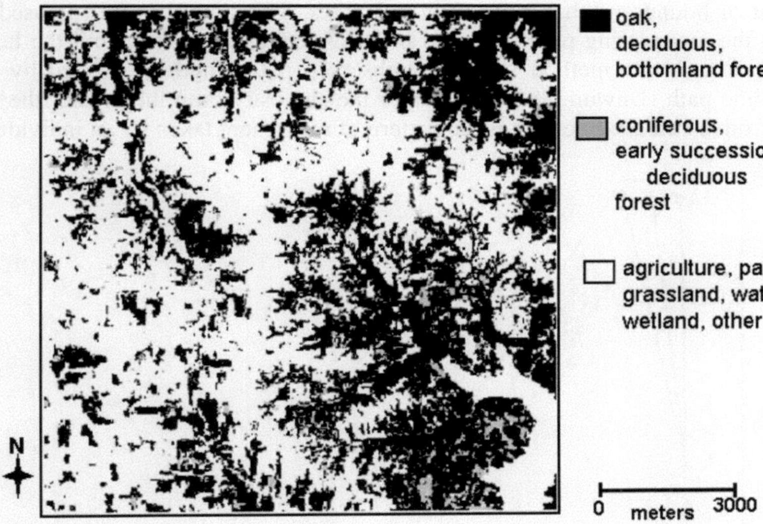

Fig. 1. This is the study area. It shows the distribution of major land cover types grouped according to a ranking as preferred habitat for gray squirrels. Black includes oak forest, mixed oak, deciduous forest, and oak/deciduous bottomland forest. Note that the majority of area colored in black is classified as oak or oak/hickory forest in the Kentucky GAP data set. The gray areas are of lesser preference and are composed primarily of coniferous forest and early succession deciduous forest. The white areas are of little habitat value to gray squirrels.

The rate at which the membership declines from β is controlled by θ. Robinson and Graniero (2005a) used a value of 0.003 for θ. Simulations will be run using $\theta = \{0.0015, 0.003, 0.006\}$. These values vary from less to greater change per unit distance. For a given β the effect is that a value of 0.0015 extends the perceptual range, but does not extend the 1.0 level beyond β. The effect of 0.006 is to constrict the perceptual range to a smaller area (Figure 3).

The manipulation of the perceptual range will affect the landscape information at each step. However, the number of steps may have a cumulative effect in that the more steps an agent is allowed to take the greater the information gathered about the landscape and increases, theoretically, the possibility of finding a suitable location. Simulations will be run for each of the fuzzy decision models for step limits of 10, 15, and 30 for the cast of $\beta = \{30, 60, 90\}$ and $\theta = 0.003$ which are the same as used by Robinson and Graniero with a step limit of 10.

There are two basic measures that may be used to describe the result of the behavior of theses agents. First, there is the state the individual reaches during the simulation. Of particular interest for each simulation is how many agents are able to reach the state of *home*. Those that fail to reach a suitable location but remain in study area are said to reach a state of *dead* while those few whose movement takes them out of bounds reach a state of *outofbounds*. The sinuosity ratio is used to describe the meandering path taken by an individual as it moves about the landscape. It is a simple method where the observed path length is divided by the straight-line path (Unwin 1981). The larger the sinuosity ratio the greater the deviation from a straight-line path is the pattern of movement taken by an individual.

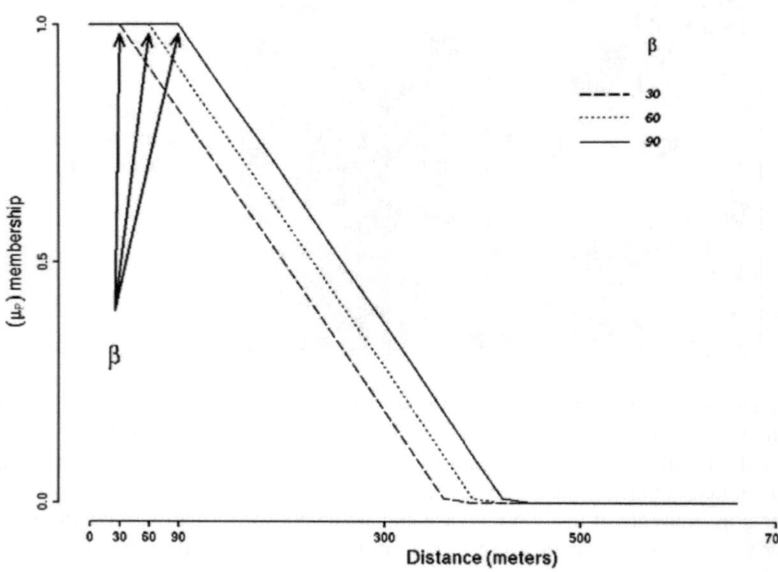

Fig. 2. Membership in the perceptual range of an individual agent as it varies with distance and the value of β.

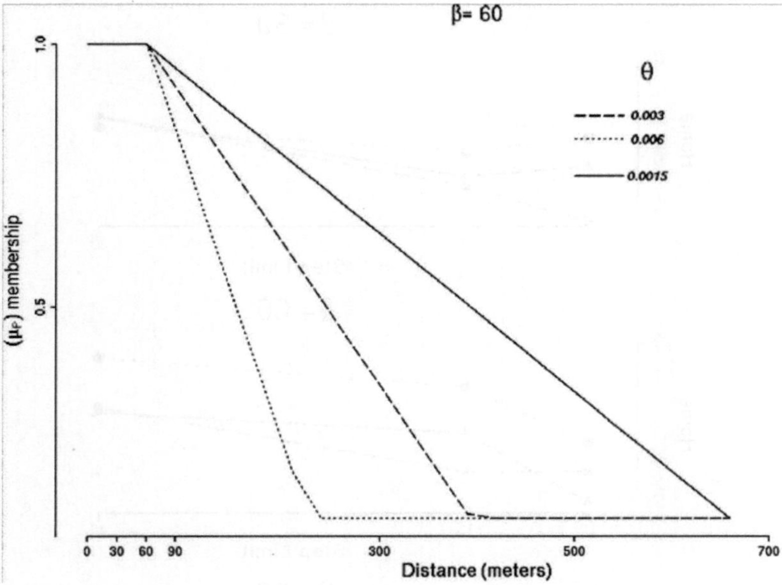

Fig. 3. Membership in the perceptual range of an individual agent as it varies with distance and the value of θ. Note that $\beta = 60$ for this example.

4 Results and Discussion

Simulations were conducted to address the three basic questions:

1. Would increasing the limit on the number of steps increase the success rate in finding a home regardless of the decision model?
2. Would variations in β and/or θ affect each of the decision models in the same manner in terms of success in finding a home?
3. Would variations in β and/or θ affect each of the decision models in the same manner in terms of the sinuosity of the path taken to find a home?

Although an increase in the number of steps generally increases the number of individual able to find a home location, its effect is not uniform between or within each of the decision models. The most linear pattern of results is associated with the largest spatial extent of ${}^{0+}P(x)$ which is associated with $\beta = 90$. Even so, it is notable that when ${}^{0+}P(x)$ is at its least the noncompensatory model is most successful when until the step limit is increased to 30. Then there is a convergence of all the decision models with the noncompensatory model being slightly less successful. It is interesting to note that the compensatory model is generally the model that is consistently the second most successful model. When the perceptual range is the most restricted (i.e.,, $\beta = \{30, 60\}$) the population of agents using the

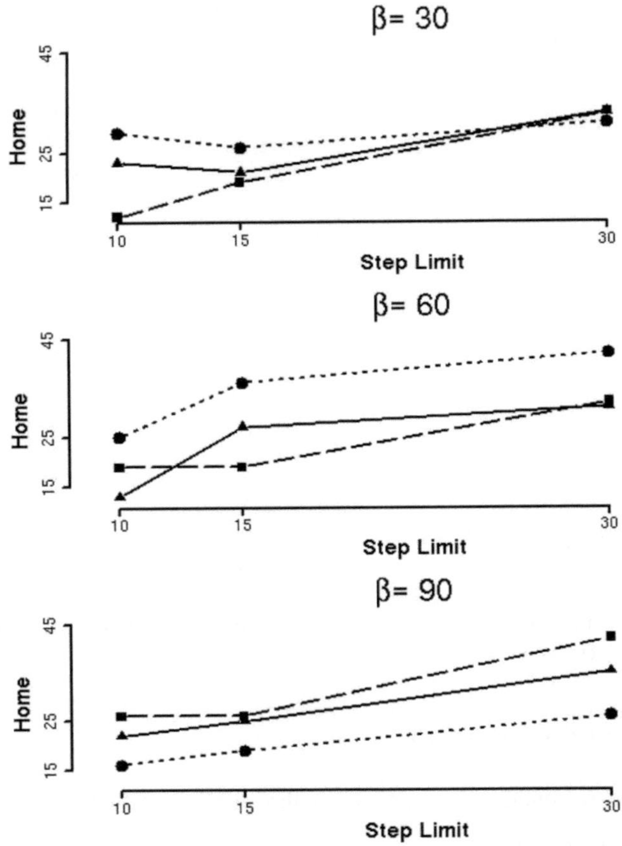

Fig. 4. Effect on home finding success in relation to increasing the step limit by the different values of β. The x-axis is the number of agents who found suitable locations for establishing a home range. The dotted lines with circles represent results for the noncompensatory fuzzy decision model population. Solid lines with triangles represent results for the compensatory fuzzy decision model population. Dashed lines and squares represent with squares represent results for the Yager decision model population. In this case $\theta = 0.003$.

noncompensatory model tend to be more successful. The results for the case of $\beta = \{30, 60\}$ also illustrate that an increase in the step limit does not always instigate a proportional increase the proportion of the population successfully dispersing to a new home range (Figure 4). In particular, the case of $\beta = 60$ seem to indicate that there would be little if any additional success in the population beyond 30 steps. Even so, only about half of the 84 agents in each population would, at best, find a suitable location for establishing a home range.

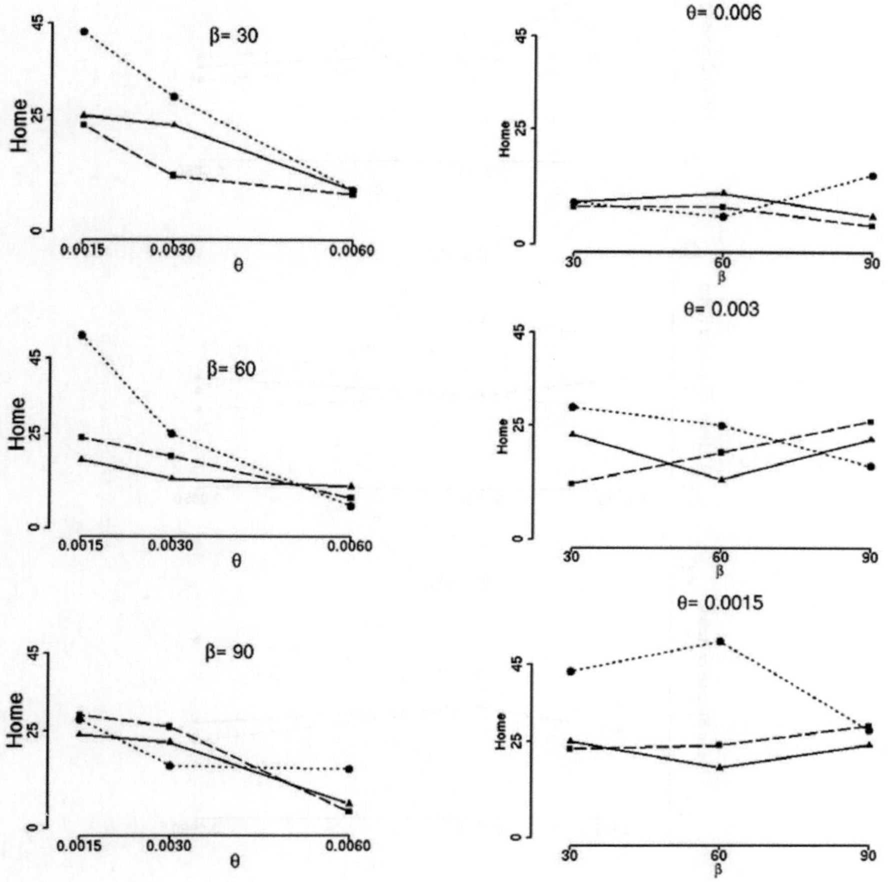

Fig. 5. Number of individual agents finding a home range as the perceptual range is varied with a combination of different values of β and θ. The *x*-axis is the number of agents who found suitable locations for establishing a home range. The plots on the left show how the results vary with changes in θ given a particular value of β. Plots on the right show how results vary with changes in β given a particular value of θ. The dotted lines with circles represent results for the noncompensatory fuzzy decision model population. Solid lines with triangles represent results for the compensatory fuzzy decision model population. Dashed lines and squares represent with squares represent results for the Yager decision model population. The step limit for all was 10.

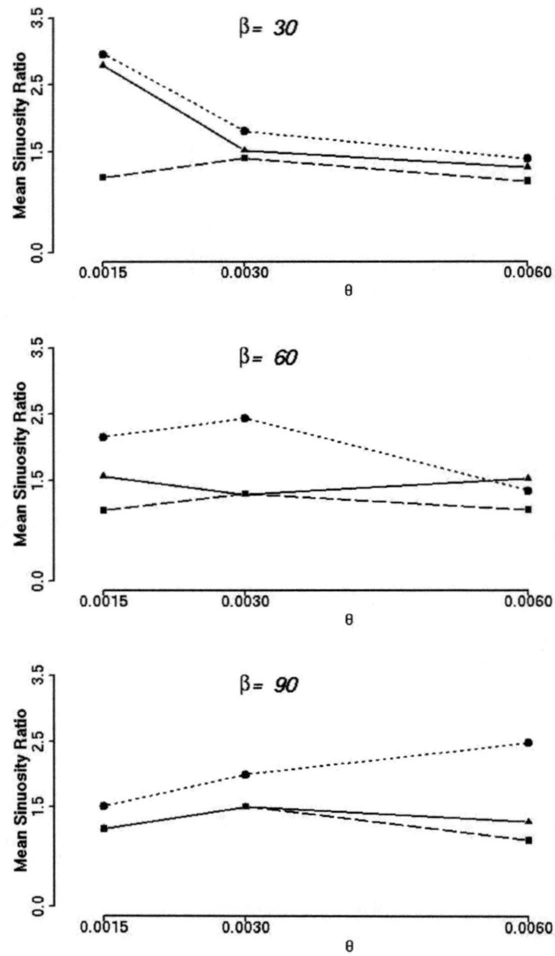

Fig. 6. The relationship between the mean sinuosity ratio of paths taken by agents that were successful in finding home range locations in relation to variations in β and θ. The dotted lines with circles represent results for the noncompensatory fuzzy decision model population. Solid lines with triangles represent results for the compensatory fuzzy decision model population. Dashed lines and squares represent with squares represent results for the Yager decision model population.

Generally speaking, given a value of β the number of agents having success at finding a suitable home range location clearly declines as ${}^{0+}P(x)$ declines. It is striking how all decision models converge to similar levels of very low success when $\theta = 0.006$ for $\beta = \{30, 60\}$. The same can be said in the case of $\beta = 90$ with the exception of the noncompensatory decision model population. Although the noncompensatory model population does not outperform the others for $\theta=\{0.0015,0.003\}$, it does for the case of $\theta=0.006$. The combination of $\beta=90$ and $\theta=0.006$ means the membership curve is approaching that of crisp set. The compensatory and Yager decision models do differ, but tend towards having results that are similar, more so than when compared with the noncompensatory decision models (Figure 5).

The sinuosity ratio provides an indication of the degree to which the agent took a meandering versus a straight-line path to its final home range location. Agents using the noncompensatory decision model, on average, took more meandering paths than those using either the compensatory or Yager decision models. This is most pronounced when $\theta=0.0015$. While pattern of results for the compensatory model are similar to the noncompensatory model for the case where $\beta=30,\theta=0.0015$. Both models have high mean sinuosity ratios that imply that the agents tended to wander about more trying to find a suitable home range location. Agents using the Yager decision model, on average, tend to meander much less in their search than do those using the noncompensatory model. In addition, the mean sinuosity ratio for the Yager decision model appears to be relatively insensitive to the variations in how the perceptual range is defined (Figure 6).

5 Concluding Comment

This cursory exercise investigating the sensitivity of agent's decisions to variations in the perceptual range shows that there are few situations where the fuzzy decision models are insensitive to changes in the perceptual range. When the membership curve approaches the shape of a crisp set (i.e., when $\theta=0.006$) then we see all decision models being relatively insensitive and approaching the dismal performance of the crisp decision model as noted in Robinson and Graniero (2005a). More often is the case that there are differences in behavioral outcomes given changes in the characteristic function of the perceptual range. There was not one fuzzy decision model that outperformed the others in all cases. The agents using the noncompensatory fuzzy model tended to be more successful when $\beta=\{30,60\},\theta=\{0.0015,0.003\}$.

In general, agent behavior did vary in relation to how the perceptual range was configured. This indicates that the perceptual range is a crucial concept in individual based models as they function much like a spatially database view limiting the query to a portion of the landscape. This study kept the landscape as a constant. However, given the importance of the characteristic function of the perceptual range in filtering informational flow to an agent, it seems reasonable to investigate the effects of varying landscape characteristics.

Acknowledgements

The partial support of a Natural Sciences and Engineering Research Council (NSERC) of Canada Discovery Grant is gratefully acknowledged.

References

Bellman, R.E., Zadeh, L.A.: Decision-making in a fuzzy environment. Management Science 17, 141–164 (1970)

Bowman, J., Jaeger, J.A.G., Fahrig, L.: Dispersal distance of mammals is proportional to home range size. Ecology 83, 2049–2055 (2002)

Graniero, P.A., Robinson, V.B.: A probe mechanism to couple spatially explicit agents and landscape models in an integrated modelling framework. International Journal of Geographical Information Science 20(9), 965–990 (2006)

Klir, G.J., Yuan, B.: Fuzzy Sets and Fuzzy Logic: Theory and Applications. Prentice-Hall, Upper Saddle River (1995)

Lima, S.L., Zollner, P.A.: Towards a behavioral ecology of ecological landscapes. Trends in Ecology and Evolution 11, 131–135 (1996)

Robinson, V.B.: Using fuzzy spatial relations to control movement behavior of mobile objects in spatially explicit ecological models. In: Matsakis, P., Sztandera, L.M. (eds.) Applying Soft Computing in Defining Spatial Relations, pp. 158–178. Physica-Verlag, Heidelberg (2002)

Robinson, V.B., Graniero, P.A.: Spatially explicit individual-based ecological modeling with mobile fuzzy agents. In: Petry, F.E., Robinson, V.B., Cobb, M.A. (eds.) Fuzzy Modeling with Spatial Information for Geographic Problems, pp. 299–334. Springer, Heidelberg (2005a)

Robinson, V.B., Graniero, P.A.: Modeling with fuzziness in the extensible component objects for constructing observable simulation models (ECO-COSM) system. In: Ma, Z. (ed.) Advances in Fuzzy Object-Oriented Databases: Modeling and Applications, pp. 269–300. Idea Group, London (2005b)

Unwin, D.: Introductory Spatial Analysis. Methuen, London (1981)

Wolff, J.O.: Behavioral model systems. In: Barrett, G.W., Peles, J.D. (eds.) Landscape Ecology of Small Mammals, pp. 11–26. Springer, New York (1999)

Fuzzy Multidimensional Databases

Anne Laurent

Abstract. The interest for *OLAP* (standing for On-Line Analytical Processing), working on multidimensional databases is growing dramatically due to its interest in data analysis and data mining. Recent works (LBMD+00),(LGM00) showed the great interest of integrating fuzzy set theory in such technologies in the framework of data mining. We now propose to enhance the multidimensional data model to handle fuzziness. This model then provides the way to apply *OLAP Mining* methods on Fuzzy Multidimensional Databases, for *Fuzzy-OLAP Mining*.

1 Introduction

Fuzzy set theory has proven to be very interesting when representing information from the real world, and to query databases. The study of fuzzy databases is an active area, and many works exist, dealing with all models, especially the relational and object-oriented ones ((BJ95)).

In this context, Dr. Ashley Morris contributed a lot to the evolution of existing Entity-Relation models to the handling of the real world by means of Fuzzy Logic (VSM02; MPC98). Especially focusing on Geographical Information, these extensions have dealt with a lot of kinds of data, as various data types are handled by Geographical Information Systems (also known as GIS).

More recently, the process of data mining has led to many works. Many of the most recent ones focus on *OLAP* (standing for On-Line Analytical Processing) and consider data stored at a particular aggregation level in *multidimensional databases*. These databases are usually built from *data warehouses* which are massive databases built from heterougeneous data sources and used for querying, reporting and analysis. Multidimensional databases provide means to deal with massive data from such data warehouses in an analytical framework. They prove their great interest for data mining when coupled with learning systems ((Han98)), and especially with fuzzy learning systems ((LBMD+00), (LGM00), (KA05)).

Anne Laurent
LIRMM - Univ. Montpellier 2 - CNRS UMR 5506, 161 rue Ada, Montpellier
e-mail: laurent@lirmm.fr

J. Kacprzyk et al. (Eds.): Uncer. Appro. for Spatial Data Model. and Process., SCI 271, pp. 43–59.
springerlink.com

It should be noted that OLAP systems are highly correlated with GIS, as they handle heterogeneous data types, and try to put them together and aggregate them through hierarchical taxonomies in order to facilitate the users' interactions with the data.

However, only a few works introduced the management of imprecision in mutidimensional databases (FD99), (PJ99; PJD99). More recently, (MSVRA06) has defined how introducing fuzziness.

Our purpose is to provide a model to represent and manage imprecise and uncertain data at any level of the model, concerning either dimensions, hierarchies, or cell values. Our model handles flexible queries and fuzziness both in the data representation and in the management of data (operations). Thus it offers perspectives for *Fuzzy-OLAP Mining,* which associates fuzzy data mining and fuzzy multidimensional databases.

The paper is organized as follows: Section 2 presents the multidimensional data model, Section 3 presents and discusses existing works concerning the introduction of fuzziness in such databases. Section 4 presents the model we designed. Finally, Section 5 concludes and gives some associated perspectives.

2 The Multidimensional Model

The amount of available data is growing dramatically, stored mainly in data warehouses. OLAP emerged in order to handle these large amounts of data for analysis processes. This framework provides means to deal with this kind of databases efficiently. Data are extracted from the data warehouse and stored in multidimensional databases.

Many models of such databases exist (VH96; AGS97; LW96; CT98; GL97; BPT97; Vas98; ABD^{+}99). They have been compared according to their main features ((VS99)). It appears that these models define the entities that constitute the

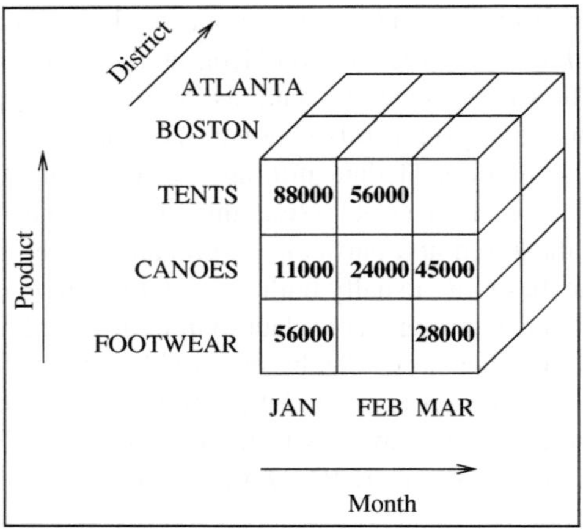

Fig. 1. An hypercube.

database (cells, dimensions and hierarchies) and algebraic operations to manipulate these entities.

Generally speaking, a multidimensional database is a set of *hypercubes* (hereafter *cubes*). A cube is designated by means of a set of dimensions. Each dimension is associated with a domain of values. Hierarchies may be defined on dimensions, and data may then be aggregated and viewed at several levels of granularity. A dimension of particular interest is chosen as the *measure* whose values are contained in the cells. A cell is described by means of its position on all dimensions. An example of such a cube is given in Fig. 1. When visualizing a cube, choices have to be done since there is no way to visualize high dimensional data. For instance, accross and down dimensions have to be chosen in order to determine the side to be displayed.

Algebraic operations on hypercubes are defined:

- **operations on presentation** modify the way the cube is visualized without changing the data themselves. These operations are *rotate*, *switch* (changes the order of the values of a dimension), *nest* ...
- **operations on data** extract subcubes by *slice and dice* and change the level of granularity by *rolling up and drilling down* cubes.
- **binary operations** refer to classical operations (*union, intersection, ...*)

The existing models are either defined as extensions of the relational model ((LW96; GL97; BPT97; GBLP96)), or directly defined as multidimensional data repositories ((AGS97; CT98; Vas98; ABD$^+$99)).

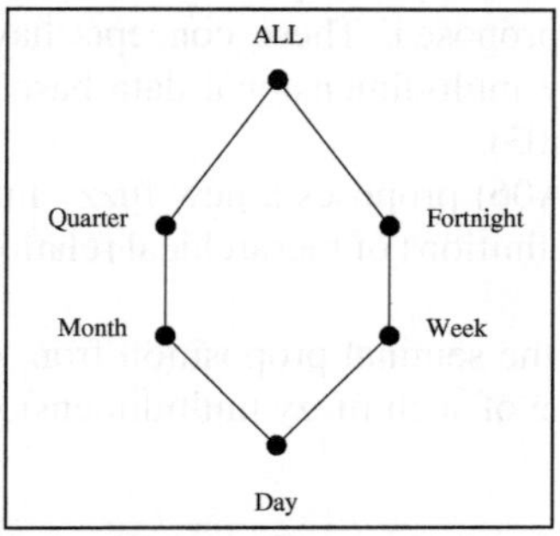

Fig. 2. A multiple hierarchy.

The hierarchies are variously defined in the litterature. (AGS97) does not explicitely define hierarchies; roll-up operations are performed considering merging functions. (BPT97) and (ABD$^+$99) consider tree-like hierarchies. But this representation does not allow *multiple* hierarchies (see Fig. 2). Thus, several other models consider partial ordering on levels to design hierarchies (*e.g.* (CT98)) with or without upper and lower bound, designing thereby a lattice or not.

Nowadays, the study of OLAP and multidimensional databases are very active areas; however only a few works introduced fuzziness.

3 Existing Works Introducing Fuzziness in the Multidimensional Model

(FD99), (PJ99; PJD99) introduced the idea of fuzziness in multidimensional databases.

(FD99) introduces means to visualize trends from the cube that are expressed in a natural way, using linguistic terms. It defines three levels for the presentation of the data to the end user. The first level consists in classical summaries (e.g. average), the second level translates these summaries by using fuzzy terms, while the third level introduces quantifiers in order to generate quantified summaries. An extension of SQL is proposed to query the data warehouse, the queries include the linguistic terms and the quantifiers that will be used in the computation of the summary. Thus this does not constitute a model of multidimensional databases, but a way to display data to the user in a more natural way by means of SQL-like queries.

In (PJ99; PJD99; JKPT04), the authors identify nine requirements they assume the models should verify and propose a model and the corresponding algebra to address all the requirements. They consider the problem of missing data for some level of information. For instance, one knows the family of diseases the patient suffers from, but not the exact one.

Uncertainty is introduced through hierarchies considering degrees between 0 and 1 in the partial order building the lattice: $e_1 \leq_p e_2 \Leftrightarrow e_1 \leq e_2\ with\ probability\ p$.

Queries are processed depending on the level of available data.

However, all these works do not handle fuzziness by means of the Fuzzy Logic framework. In (LBMD+00), the primary definition of coupling Fuzzy logic and OLAP systems has been proposed. These concepts have been extended for the definition of the first fuzzy multidimensional data base model (Lau02b; Lau02a; LBMD02; LBMD+00; Lau03).

More recently, (MSVRA06) proposes a new fuzzy multidimensional model by extending the traditional definitions of hierarchical relationships and imprecise data cubes.

In this paper, we detail the seminal proposition from (Lau02a) and extend it to the study of the construction of such fuzzy multidimensional databases.

4 Proposed Fuzzy Multidimensional Model

Fuzziness can be introduced and handled at different levels. The data contained in the database can indeed be crisp or fuzzy due to the imperfection in their knowledge. The need for transformation into fuzzy values and their manipulation is important since it leads to more general and understandable knowledge for the users.

Thus, we introduce a model (Lau01) still handling crisp data, but also handling fuzziness for data storage (section 4.1) and data manipulation (section 4.2). First of all, we define the seminal entities constituing a fuzzy hypercube (or fuzzy cube), which are *elements*, dimensions, and hierarchies, and we propose the definitions for operations to manipulate these fuzzy entities.

4.1 Data Representation

In this part, we introduce a formalization of the different kinds of entities constituting a cube and we propose a formalization for hierarchies.

4.1.1 Elements

Data in cells and dimensions may be imprecise and/or uncertain. Such data are called *elements*.

Given a reference set X, $\mathsf{F}(X)$ denotes the set of fuzzy sets of X, including singletons of X.

Definition 1. *A* value *v on a reference set X belongs to* $\mathsf{F}(X)$ *and is:* **(i)** *either a crisp value from a classical set (for instance a real value from* $X = \mathfrak{R}$*),* **(ii)** *or an imprecise value represented by its membership function.*

Definition 2. *Given a reference set X, an* element *e is defined by the couple* (v, d) $\in \mathsf{F}(X) \times [0,1]$ *where:*

- *v is a* value,
- $d \in [0,1]$ *is the* degree *of confidence attached to the value v.*

4.1.2 Hierarchies

The multidimensional model and OLAP analysis are designed to deal with hierarchies. These hierarchies are defined in different ways in existing crisp models (as described in section 2). In our approach, both the data and the hierarchies may be imprecise and uncertain. Thus we define two types of hierarchies (see Fig. 4 and 5). On the one hand these hierarchies may be defined by means of fuzzy partitions of the universe, and on the other hand, they may be defined by means of fuzzy relations.

Hierarchies of Fuzzy Partitions

Fuzzy partitions are variously defined in the litterature. In our context, fuzzy partitions are defined with the constraint of summing to 1 on all the universe:

Definition 3 (Fuzzy Partition). *A* fuzzy partition *P on the universe X is a family of fuzzy subsets* $\{F_1, ..., F_L\}$ *with membership functions* $f_1, \ldots, f_L$ *so that for all* $x \in X$, $\Sigma_{i=1}^{L} f_i(x) = 1$.

An relation $\prec$ is defined on partitions to construct hierarchies.

Definition 4. *A partition* $P_\alpha = \{F_1, \ldots, F_{L_\alpha}\}$ *preceeds a partition* $P_\beta = \{F'_1, \ldots, F'_{L_\beta}\}$ *if* $P_\alpha \neq P_\beta$ *and for all* $F' \in P_\beta$*, there exists an interval* $I \subset [1, L_\alpha]$ *so that* $F' = \bigcup_{i \in I} F_i$ *(where* $\bigcup$ *is an union operator of fuzzy subsets). The corresponding notation is* $P_\alpha \prec P_\beta$.

The Lukasiewicz t-conorm can be used for the union of the fuzzy subsets in order to consider only hierarchies that are intuitively correct (Fig. 4) and not counter-intuitive hierarchies as Fig. 3.

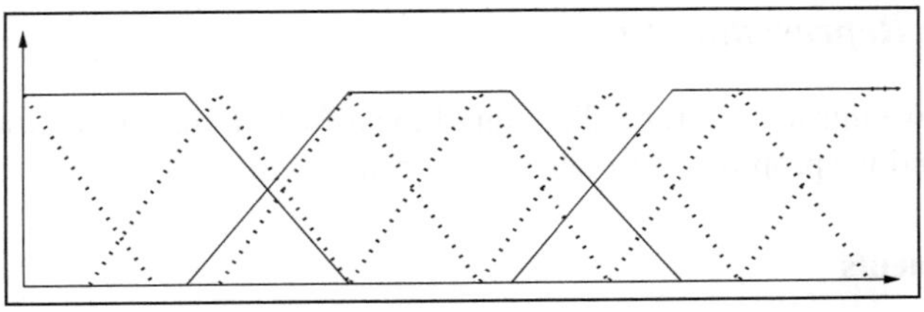

Fig. 3. Counter-Intuitive Fuzzy Hierarchy.

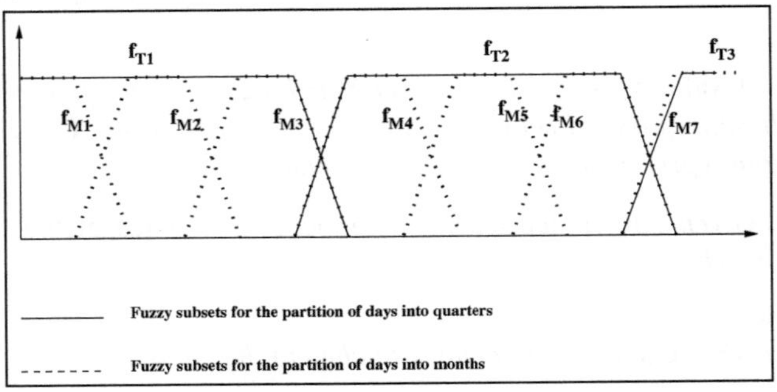

Fig. 4. Fuzzy Partition as a Hierarchy.

Given a relation $\prec$, the immediate successor of one partition in the order is defined as follows:

Definition 5. *Given* $\mathsf{P} = \{P_1, \dots, P_J\}$ *a set of J partitions, a partition* $P_b \in \mathsf{P}$ *is said to be* immediate successor *of partition* $P_a \in \mathsf{P}$ *if* $P_a \prec P_b$ *and there does not exist* $P_c \in \mathsf{P}$ *so that* $P_a \prec P_c \prec P_b$.

Simple hierarchies are distinguished from multiple ones. In our model, multiple hierarchies are taken into account by defining several simple hierarchies, so that aggregation does not take values several times into account.

Definition 6. *A fuzzy hierarchy defined by a relation* $\prec$ *on* P *is said to be* simple *if each partition* $P_i \in \mathsf{P}$ *has at most one immediate successor.*

Definition 7. *A fuzzy hierarchy defined by a relation* $\prec$ *on* P *is said to be* multiple *if there exists at least one partition* $P_i \in \mathsf{P}$ *having more than one immediate successor.*

Hierarchies defined by means of Fuzzy Relations

Now, we consider fuzzy relations (Fig. 5). Hierarchies are then defined by fuzzy strict order relations ((KY95)) antireflexive and transitive.

The transitive closure relation is computed to obtain all degrees associating any value with any other one, regardless of the size of the path.

Definition 8. *Calling G the graph associated with a fuzzy strict order relation R with membership function* f_R*, we call* transitive closure *the relation* R_T *so that* $f_{R_T}(x,y) \neq 0$ *iff there exists a path in G from x to y.*

If R_T *is the transitive closure of R, then we compute* R_T *with the following algorithm:*

1. $R' = R$

2. While R' *is modified*

$R = R'$

$R' = R \bigcup (R \circ R)$

3. Stop. $R_T = R'$

For hierarchies defined by means of relations also, we distinguish simple hierarchies from multiple ones:

Definition 9. *A hierarchy defined on a set of values V is said to be* simple *if it is defined by a fuzzy strict order relation R so that for all* $x \in V$ *there exists at most one* $b \in V$ *so that* $f_R(x,b) = 1$.

Definition 10. *A hierarchy defined on a set of values V is said to be* multiple *if it is defined by a fuzzy strict order relation so that there exists at least* $(x,y,z) \in V^3$ *so that* $f_R(x,y) = f_R(x,z) = 1$.

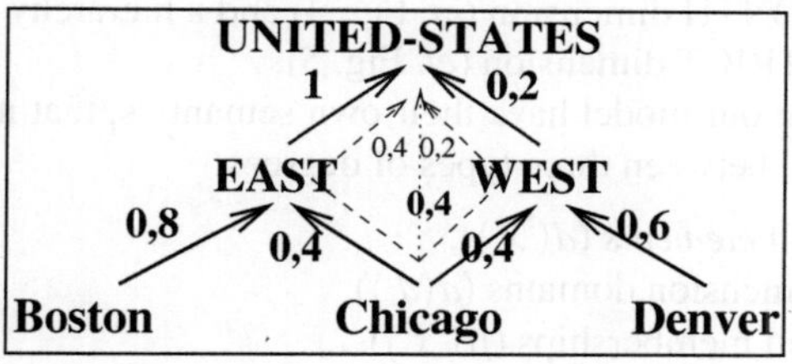

Fig. 5. Fuzzy relation as a hierarchy.

4.1.3 Generalized Dimensions

Dimensions are defined on domains and structured with hierarchies.

Definition 11. *A* domain *dom on universe X is a finite set of* elements.
V_{dom} *denotes the set of values from the* elements *of such a domain dom.*

A generalized dimension is defined as follows:

Definition 12. *A generalized dimension D is a tuple* (n, X, dom, H_P, H_R) *where:*

- *n is the name of the dimension,*
- *X its reference universe,*
- *dom its domain,*
- H_P *a set of simple hierarchies defined by an order relation* ($\prec$) *on fuzzy partitions on the domain,*
- H_R *a set of simple hierarchies defined by fuzzy strict order relations.*

We denote by D the set of all dimensions D_i in the multidimensional database.

4.1.4 Fuzzy Cubes

Definition 13 (Fuzzy Cube). *A fuzzy cube is a relation C:* $D_1 \times \ldots \times D_k \rightarrow D_C \times [0,1]$ *where* $D_i \in \mathsf{D}(i = 1,...,k)$*, and* $D_C \in \mathsf{D}$ *is the measure.*

Each cell $\overrightarrow{x}$ of the cube C is associated with the *element* $(v_C(\overrightarrow{x}), d_C(\overrightarrow{x}))$ and a value of a dimension is associated with the element $(v(d_i), d(d_i))$ (where d_i belongs to the domain of the dimension). Thereby, the degrees $d(d_i)$ correspond to the extent to which each slice d_i belongs to the cube.

A degree $\mu(\overrightarrow{x})$ is attached to each cell $\overrightarrow{x}$ to indicate to which extent $\overrightarrow{x}$ belongs to the cube.

A fuzzy cube is a classical cube if for all $i = 1,\ldots,k$, for all d_i in dom_i, $d(d_i) = 1$ and for all $\overrightarrow{x}$, $\mu(\overrightarrow{x}) = 1$ and $v(\overrightarrow{x})$ is precise.

4.1.5 Example of Fuzzy Cube

Fig. 6 shows an example of a fuzzy cube. In this example, sales of product *TENTS* in *CHICAGO* for the month of *JANUARY* have been *bad*. This value is known with confidence 0.8 and this cell belongs completly to the cube *μ(TENTS,CHICAGO,JANUARY)* $= 1$. There exists a hierarchy defined by fuzzy partitions on the MONTH dimension (*cf.* Fig. 4) and a hierarchy defined by a fuzzy relation on the DISTRICT dimension (*cf.* Fig. 5).

All the degrees in our model have their own semantics, that make them all relevant. We distinguish between three types of degrees:

- the degrees in cell *elements* $(d(\overrightarrow{x}))$,
- the degrees in dimension domains $(d(d_i))$,
- the degrees of cell memberships $(\mu(\overrightarrow{x}))$.

The first type of degree $d(\overrightarrow{x})$ represents the confidence associated with the corresponding $v(\overrightarrow{x})$ value, while the second one represents the extent to which slices belong to the cube. The third one is associated with the whole information given by the cell position (elements of the domains of all dimensions) and the cell *element* itself $((v(\overrightarrow{x}), d(\overrightarrow{x})))$. It represents the extent to which this whole information belongs to the cube. In the *SALES* cube from the previous example (Fig. 6), the μ

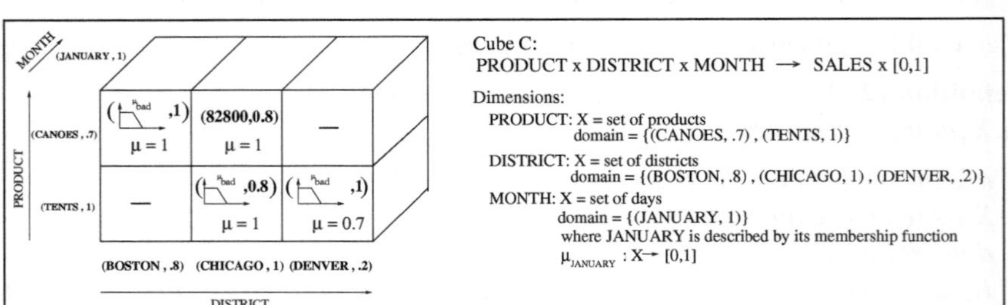

Fig. 6. A fuzzy cube.

degrees are associated with the whole information given by the *district*, *product*, *month*, and *sales* values, and not only with the *sales* value, while the $d(\overrightarrow{x})$ degrees are associated only with the corresponding $v(\overrightarrow{x})$ values reprensenting the sale rates.

Thus these degrees may come from the data themselves. For instance, one may build a cube from various data sources. In this case, these different degrees offer the opportunity of distinguishing between the degree of fiability of the data source (which will be represented by the μ degrees) and the cell data themselves, whose uncertainty will be represented by the $d(\overrightarrow{x})$ degrees.

But the degrees may also have been introduced by operations when querying the multidimensional database, as we will explain in the next section (4.2). These degrees will then represent the extent to which entities (cells or slices) belong to the resulting cube. Many fuzzy relational models have focused on fuzzy queries, and do not represent the result of these queries in their model. In our model, we define a close algebra, where operations applied on fuzzy cubes result in fuzzy cubes represented in the same model. This appears to be very important since several operations may then be applied one after the other.

4.2 *Operations*

Our model allows the representation of imperfect data in fuzzy cubes. We define now the operations to manipulate these cubes.

We need comparison measures between fuzzy subsets to determine to which extent two fuzzy values are similar. In our case, we will have to evaluate the *satisfiability* to a reference description (a fuzzy criterion described by its membership function) by the description of the value to compare, which is a particular kind of measure of resemblance. Thus, the needed measure must enable the measurement of the inclusion of a description into a reference. Such measures are called *measures of satisfiability* (BMRB96). We use the notation S for these operations. In our model, we use a measure verifying the following property in the case of a precise value: $if\, B = \{b\}\, then\, \mathsf{S}(A,B) = \mu_A(b)$.

4.2.1 Operations on Presentation

In section 4, we have presented the operations associated with the vizualisation of the data by the user. Since they are not modified when we consider fuzzy data, we focus on the two other kinds of operations.

4.2.2 Operations Affecting Data (*Intra-Cubes* Operations)

Selection on cell values

The *Slice* OLAP Operation consists in selecting the cells of a cube that satisfy a given selection criterion (Fig. 7).

In crisp models, such an operation results in a cube with the same dimensions as previously, in which cell values have been either unchanged or put to null depending on whether they satisfy or not the selection criterion.

In our model, both the selection criterion and the cell values may be fuzzy. The resulting cube has also the same dimensions as the previous one. All cell *elements* are unchanged (for each cell $\overrightarrow{x}$, $v_C(\overrightarrow{x})$ and $d_C(\overrightarrow{x})$ are unchanged). But new cube membership degrees μ are computed, and the cells belong to the resulting cube with a degree depending on the degree of satisfiability of $v_C(\overrightarrow{x})$ to the given selection criterion, on the degree of confidence in the value of the cell, and on the former cube membership degree.

To combine these values, we use an operator $\top$. This operator may be a t-norm $\top$. But other operations may be considered, such as for instance an operation which would consider differently the degrees to merge depending on their semantics (Det00). But this study is beyond the scope of this paper. In the remainding part of this paper, we will use the notation $\top$ to refer to this merging operator.

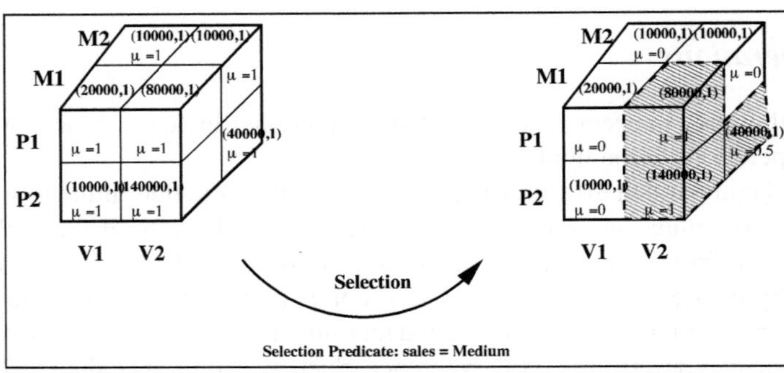

Fig. 7. Selection on cell values.

Given a fuzzy cube $C : D_1 \times \ldots \times D_k \to D_C \times [0,1]$, the resulting cube C' from a selection operation on cell values by the fuzzy criterion O with membership function μ_O is defined as: $C' : D_1 \times \ldots \times D_k \to D_C \times [0,1]$. For each dimension D_i, $i = 1,\ldots,k$, $dom_i(C') = dom_i(C)$. Denoting $\overrightarrow{X} = dom_1 \times \ldots \times dom_k$, for all $\overrightarrow{x} \in \overrightarrow{X}$, $v_{C'}(\overrightarrow{x})$=$v_C(\overrightarrow{x})$ and $d_{C'}(\overrightarrow{x})$=$d_C(\overrightarrow{x})$.

The new degrees $\mu_{C'}$ are computed using the degrees of satifiability $\mathsf{S}(\mu_O, v_C(\overrightarrow{x}))$, the degrees of confidence $d_C(\overrightarrow{x})$ and the degrees $\mu_C(\overrightarrow{x})$:
$\forall \overrightarrow{x} \in \overrightarrow{X}, \mu_{C'}(\overrightarrow{x}) = \top(\mathsf{S}(\mu_O, v_C(\overrightarrow{x})), d_C(\overrightarrow{x}), \mu_C(\overrightarrow{x}))$

Selection on dimensions values

The OLAP operation *Dice* reduces the domain of dimensions (Fig. 8).

In crisp models, a selection on dimension values reduces the domain of a dimension. Values in the domains are indeed kept or not depending on whether they satisfy or not the selection criterion.

In our model, *elements* on dimensions may belong gradually to a domain with a degree. This degree corresponds to the $d(d_i)$ value if we consider the *element* d_i. It represents the extent to which the slice corresponding to this value belongs to the cube. Thus this operation modifies this degree, taking into account the former d_i degree, and the degree of satisfiability of the selection criterion by the value of the *element.*

Let C' be the cube resulting from a selection operation on dimension D_i with domain dom_i in cube C considering a fuzzy selection criterion O described by its membership function μ_O. Each degree d_i of *elements* in domain dom_i of the concerned dimension is modified as follows:

$$\forall d_i \in dom_i, d_{C'}(d_i) = \mathsf{T}(\mathsf{S}(\mu_O, v_C(d_i)), d_C(d_i))$$

where T is an operator which may be for instance a t-norm.

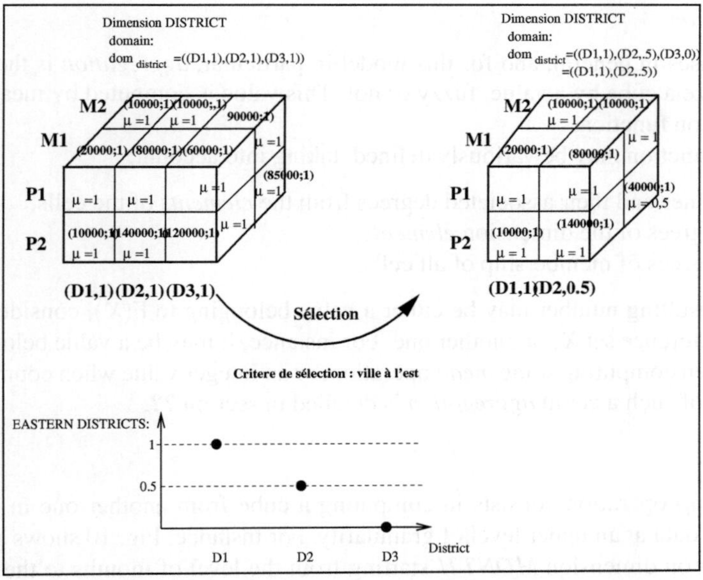

Fig. 8. Selection on dimensions values.

Projection
This operation reduces the number of dimensions of the cube (Fig. 9).

Projecting the fuzzy cube defined by the relation $C : D_1 \times \ldots \times D_k \rightarrow D_C \times [0,1]$ on dimensions $D_m, \ldots, D_n(\{D_m, \ldots, D_n\} \subseteq \{D_1, \ldots, D_k\})$ we obtain the fuzzy cube $C' : D_m \times \ldots \times D_n \rightarrow D_C \times [0,1]$

To perform this operation, a fusion operator is needed in order to reduce the domain of the dimension to erase to a singleton. In our model, this particular fusion operation is seen as a prelimary operation before projecting the cube. It may be performed by *rolling up* the cube (see following sections).

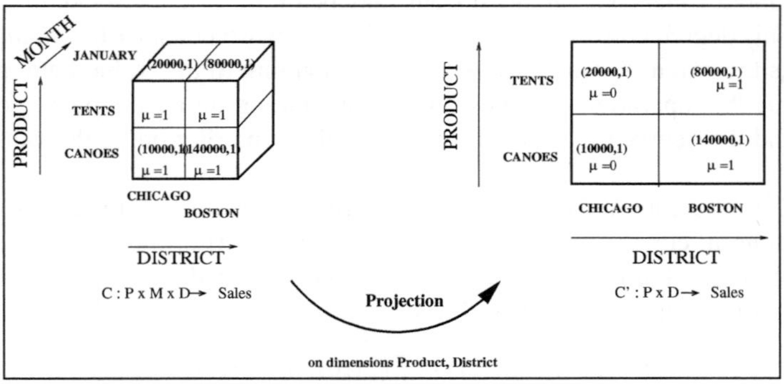

Fig. 9. Projection.

Aggregation

In databases in general, and for this model in particular, *aggregation* is the way to summarize a cube by a value, fuzzy or not. This value is computed by means of an aggregation function.

This function may be variously defined, taking into account:

- the values and their associated degrees from the *elements* of the cells,
- the degrees of the dimension *elements*,
- the degrees of membership of all cells.

The resulting number may be either a value belonging to $\mathsf{F}(X)$, considering the former reference set X, or another one. For instance, it may be a value belonging to $\mathsf{F}(X)$ when computing some *mean* operation, or an integer value when counting. An example of such a count *aggregation* is detailed in section **??**.

Roll-Up

The *roll-up* operation consists in computing a cube from another one in order to visualize data at an upper level of granularity. For instance, Fig. 10 shows a *roll-up* operation on dimension $MONTH$ starting from the level of months to the level of quarters.

This operation requires one dimension to be chosen. And since several hierarchies may be defined on a dimension, the chosen hierarchy has to be known. This hierarchy may be either defined by means of partitions and an associated $\prec$ relation, or by means of a fuzzy strict order relation. In crisp models, hierarchies are crisp, and each cell value is taken into account for one particular value in the upper level of granularity. This value may for instance participate for the calculus of the mean, for counting etc.

In our model, cell values may participate for several values in the upper level due to their fuzziness. For instance, as we saw in figure 5, when considering the $DISTRICT$ dimension, $CHICAGO$ is taken into account for both $EAST$ and $WEST$ when rolling up from the city level to the region level. considering the temporal

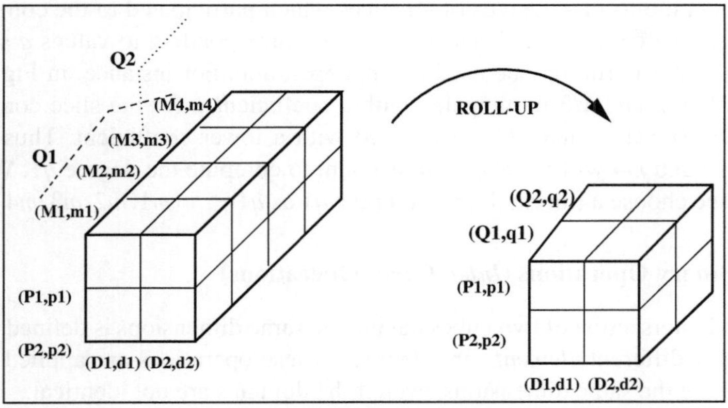

Fig. 10. Roll-Up.

dimension, the first four months would participate in building the value for the first quarter (see Fig 4).

The degree indicating at which extent a cell has to be taken into account when rolling up to an upper value is variously computed, depending on the nature of the hierarchy.

We consider a dimension having domain *dom* and values V_{dom}, which is the current state before rolling up. $V_{dom'}$ is the set of values for the resulting upper level in cube C'. We write $c(a,b)$ the coefficients indicating to which extent each value $b \in V_{dom'}$ is reachable from values $a \in V_{dom}$. These coefficients are variously computed depending on the hierarchy type the user chooses to roll up the cube:

- hierarchy $h_1 \in H_R$ defined by means of a relation R:
  First, the transitive closure R_T of R is computed. Then $c(a,b) = f_{R_T}(a,b)$ indicates to which extent b is reachable from a.
- hierarchy $h_2 \in H_P$ defined by means of two partitions P_α and P_β ($P_\alpha \prec P_\beta$):
  $c(a,b) = f_b^\beta(a)$ indicates to which extent b is reachable from a, where f_b^β is the membership function describing the b value in the family of fuzzy sets which constitutes the partition P_β.

These coefficients are taken into account when computing the new fuzzy cube.

For all $b \in V_{dom'}$, the elements of the new cubes $C'(d_1,\ldots,b,\ldots,d_k)$ are calculated with a *roll-up* relation which takes into account:

- the coefficients $c(j,b)$ ($j \in V_{dom}, b \in V_{dom'}$),
- the values and their associated degrees from the *elements* of the cells,
- the degrees of the dimension *elements*,
- the degrees of membership of all cells.

This function is used to compute the values to be stored in cells of the resulting cube. In this resulting cube, the degrees of the slices on the rolled up dimension are computed from the former degrees. For this, we consider an operation which takes into account the former degrees of all slices which participated to the computation of the resulting cell. These slices are the ones corresponding to values a such that $c(a,b) > 0$. A t-norm may be used for this operation. For instance, in Fig. 10, the values $M1$, $M2$ and $M3$ participated with a coefficient 1 for the slice corresponding to the $Q1$ value, and $M4$ participated with a lower coefficient. Thus degrees $m1, m2, m3$ and $m4$ will be taken into account to compute the degree $q1$. We could for instance choose a t-norm $\top$ and compute $q1$ as $q1 = \top(m1, m2, m3, m4)$.

4.2.3 Binary Operations (*Inter-Cubes* Operations)

Union and intersection of two cubes having the same dimensions is defined by combining their different *elements* and degrees. These operations are applied only on cubes having the same dimensions, even if the domains are not identical.

Definition 14. *Given two cubes C_α and C_β defined on $D_1^\alpha \times \ldots \times D_k^\alpha \to D_C^\alpha \times [0,1]$ and $D_1^\beta \times \ldots \times D_l^\beta \to D_C^\beta \times [0,1]$ with domains dom_i^α, $i = 1,\ldots,k$, and dom_j^β, $j = 1,\ldots,l$, these cubes have the* same structure *if:*

1. *$k = l$*
2. *for all $i = 1,\ldots,k$, the name of the dimension D_i^α is the same as the one for dimension D_i^β*
3. *D_C^α and D_C^β are dimensions whose domains are defined on the same universe.*

Union

The values of the dimension domains for the new cube are obtained considering the union of the domain values for the two cubes. If the value is only in one of the two cubes, then the degree is taken from this cube. Otherwise, the cell value is computed by the union of the two fuzzy subsets constituting the cell values of the two cubes. The degrees $d(\overrightarrow{x})$ and $\mu(\overrightarrow{x})$ are combined using an operator which may be for instance a t-conorm $\perp$. As we highlighted previously when considering the $\top$ operator, this operator may be replaced by another one.

Given two fuzzy cubes C_α and C_β having the same structure ($D_1 \times \ldots \times D_k \to D_C \times [0,1]$), the cube C_γ resulting from the union $C_\gamma = C_\alpha \bigcup C_\beta$ is defined on $D_1 \times \ldots \times D_k$ by:

$$\forall i = 1,...,k,\ V_{Dom_i}^\gamma = V_{Dom_i}^\alpha \bigcup V_{Dom_i}^\beta,$$
$$\forall \overrightarrow{x} \in \overrightarrow{X},\ v_\gamma(\overrightarrow{x}) = \cup(v_\alpha(\overrightarrow{x}), v_\beta(\overrightarrow{x})),$$
$$\forall \overrightarrow{x} \in \overrightarrow{X},\ d_\gamma(\overrightarrow{x}) = \perp(d_\alpha(\overrightarrow{x}), d_\beta(\overrightarrow{x})),$$
$$\forall \overrightarrow{x} \in \overrightarrow{X},\ \mu_\gamma(\overrightarrow{x}) = \perp(\mu_\alpha(\overrightarrow{x}), \mu_\beta(\overrightarrow{x}))$$

where $\cup$ is an union operator on fuzzy subsets.

Intersection

When computing the intersection of two fuzzy cubes, the values in domains of dimensions for the resulting cube are obtained by computing the fuzzy intersection between the two fuzzy subsets in the cubes and the degrees are calculated by combining the former degrees by means of an operator T which may be for instance a t-norm. When the value is only in one of the two cubes, it will not be in the resulting cube.

If $C_\gamma = C_\alpha \bigcap C_\beta$ is the intersection cube then it is defined on $D_1 \times \ldots \times D_k$ by:

$\forall i = 1, ..., k, V^{\gamma}_{Dom} = V^{\alpha}_{Dom} \bigcap V^{\beta}_{Dom}$,
$\forall \overrightarrow{x} \in \overrightarrow{X}, v_\gamma(\overrightarrow{x}) = \cap(v_\alpha(\overrightarrow{x}), v_\beta(\overrightarrow{x}))$,
$\forall \overrightarrow{x} \in \overrightarrow{X}, d_\gamma(\overrightarrow{x}) = \mathsf{T}(d_\alpha(\overrightarrow{x}), d_\beta(\overrightarrow{x}))$,
$\forall \overrightarrow{x} \in \overrightarrow{X}, \mu_\gamma(\overrightarrow{x}) = \mathsf{T}(\mu_\alpha(\overrightarrow{x}), \mu_\beta(\overrightarrow{x}))$

where $\cap$ is an intersection operator on fuzzy subsets.

5 Conclusion

In this paper, we describe an approach for fuzzy multidimensional databases. The general model dealing with imperfect data in multidimensional databases is presented. Moreover, the associated architecture is given in order to use these kinds of data repository for relevant knowledge discovery from large databases from the real world. According to several works that highlighted the great interest of the OLAP framework in the knowledge discovery process, this approach enhances the existing solutions.

As studied by Dr. Ashley Morris, managing fuzziness in information systems is of great interest, and leads to enhanced systems. In the framework of multidimensional databases, it not only provides the way to deal with data from the real world, and to apply flexible operations on data sets stored as multidimensional arrays, but it also provides means to generate more understandable fuzzy rules, as shown in (Lau02a).

Finally, it should be noticed that time is of one the main dimensions in data cubes, and should thus be considered in a special way, as done for generating multidimensional sequential patterns (PCL+05?). Mixing fuzzy sequential patterns (FLT07) and multidimensional sequential patterns will thus be one of the future directions.

References

[ABD+99] Albrecht, J., Bauer, A., Deyerling, O., Guenzel, H., Huemmer, W., Lehner, W., Schlesinger, L.: Management of multidimensional aggregates for efficient online analytical processing. In: Int. Database Engineering and Applications Symposium, IDEAS 1999 (1999)

[AGS97] Agrawal, Gupta, Sarawagi: Modeling multidimensional databases. In: Proc. of the 13th Int. Conference on Data Engineering (1997)

[BJ95] Bosc, P., Kacprzyk, J. (eds.): Fuzziness in Database Management Systems. Studies in fuzziness. Springer, Heidelberg (1995)

[BMRB96] Bouchon-Meunier, B., Rifqi, M., Bothorel, S.: Towards general measures of comparison of objects. Fuzzy Sets and Systems 84, 143–153 (1996)

[BPT97] Baralis, E., Paraboschi, S., Teniente, E.: Materialized view selection in a multidimensional database. In: 23rd VLDB Conference (1997)

[CT98] Cabibbo, L., Torlone, R.: A logical approach to multidimensional databases. In: Schek, H.-J., Saltor, F., Ramos, I., Alonso, G. (eds.) EDBT 1998. LNCS, vol. 1377, pp. 183–197. Springer, Heidelberg (1998)

[Det00] Detyniecki, M.: Mathematical Aggregation Operators and their Application to Video Querying. PhD thesis, University Paris 6 (2000)

[FD99] Feng, L., Dillon, T.: Enhancing data warehousing with fuzzy technology. In: Bench-Capon, T.J.M., Soda, G., Tjoa, A.M. (eds.) DEXA 1999. LNCS, vol. 1677, pp. 872–881. Springer, Heidelberg (1999)

[FLT07] Fiot, C., Laurent, A., Teisseire, M.: From crispness to fuzziness: Three algorithms for soft sequential pattern mining. IEEE Transactions on Fuzzy Systems 15 (2007)

[GBLP96] Gray, J., Bosworth, A., Layman, A., Pirahesh, H.: Data cube: A relational aggregation operator generalizing group-by, cross-tabs, and sub-totals. In: Proc. of ICDE 1996 (1996)

[GL97] Gyssens, M., Lakshmanan, V.S.: A foundation for multidimensional databases. In: Proc. 23rd Int. Conf. on Very Large Data Bases, pp. 106–115 (1997)

[Han98] Han, J.: Towards on-line analytical mining in large databases. ACM SIGMOD Record 27, 97–107 (1998)

[JKPT04] Jensen, C., Kligys, A., Pedersen, T.B., Timko, I.: Multimendional data modeling for location-based services. VLDB Journal 13, 1–21 (2004)

[KA05] Kaya, M., Alhajj, R.: Fuzzy OLAP Association Rules Mining-Based Modular Reinforcement Learning Approach for Multiagent Systems. IEEE ransactions on Systems, Man, and Cybernetics 35, 326–338 (2005)

[KY95] Klir, G.J., Yuang, B.: Fuzzy sets and Fuzzy Logic. Prentice Hall International, Englewood Cliffs (1995)

[Lau01] Laurent, A.: De l'olap mining au f-olap mining. Journées d'Extraction et Gestion des Connaissances (2001)

[Lau02a] Laurent, A.: A new Approach for the Generation of Fuzzy Summaries based on Fuzzy Multidimensional Databases. International Journal of Intelligent Data Analysis 7(2) (2002)

[Lau02b] Laurent, A.: Bases de données multidimensionnelles floues et leur application à la fouille de données. PhD thesis, Université Paris 6 (2002)

[Lau03] Laurent, A.: Querying fuzzy multidimensional databases: Unary operators and their properties. International Journal of Uncertainty, Fuzziness and Knowledge-Based Systems 11, 31–46 (2003)

[LBMD+00] Laurent, A., Bouchon-Meunier, B., Doucet, A., Gancarski, S., Marsala, C.: Fuzzy data mining from multidimensional databases. In: Quo Vadis Computational Intelligence? Studies in Fuzziness and Soft Computing, Proc. of ISCI, vol. 54 (2000)

[LBMD02] Laurent, A., Bouchon-Meunier, B., Doucet, A.: Flexible unary multidimensional queries and their combinations. In: Proc. of the 9th International Conference on Information Processing and Management of Uncertainty in Knowledge-Based Systems (IPMU 2002), pp. 315–322 (2002)

[LGM00] Laurent, A., Gançarski, S., Marsala, C.: Coopération entre un système d'extraction de connaissances floues et un système de gestion de bases de données multidimensionnelles. In: Proc. of LFA 2000, pp. 325–332 (2000)

[LW96] Li, C., Wang, X.S.: A data model for supporting on-line analytical processing. In: Proceedings Conference on Information and Knowledge Management, pp. 81–88 (1996)

[MPC98] Morris, A., Petry, F., Cobb, M.: Incorporating spatial data into the fuzzy object oriented data model. In: Proceedings of the Seventh International Conference on Information Processing and Management of Uncertainty in Knowledge Based Systems (IPMU), pp. 604–611 (1998)

[MSVRA06] Molina, C., Sánchez, D., Vila, M.A., Rodríguez-Ariza, L.: A New Fuzzy Multidimensional Model. IEEE Transactions on Fuzzy Systems 14, 897–912 (2006)

[PCL$^+$05] Plantevit, M., Choong, Y.W., Laurent, A., Laurent, D., Teisseire, M.: M^{2}SP: Mining sequential patterns among several dimensions. In: Jorge, A.M., Torgo, L., Brazdil, P.B., Camacho, R., Gama, J. (eds.) PKDD 2005. LNCS (LNAI), vol. 3721, pp. 205–216. Springer, Heidelberg (2005)

[PJ99] Pedersen, T.B., Jensen, C.S.: Multidimensional data modeling for complex data. In: Proceedings of ICDE 1999 (March 1999)

[PJD99] Pedersen, T.B., Jensen, C.S., Dyreson, C.E.: Supporting imprecision in multidimensional databases using granularities. In: 11th Int. Conf. on Scientific and Statistical Database Management, pp. 90–101 (1999)

[PLT09] Plantevit, M., Laurent, A., Teisseire, M.: Olap-sequential mining: Summarizing trends from historical multidimensional data using closed multidimensional sequential patterns. Annals of Information Systems, special issue in New Trends in Data Warehousing and Data Analysis (to appear, 2009)

[Vas98] Vassiliadis, P.: Modeling databases, cubes and cube operations. In: 10th SSDBM Conference (1998)

[VH96] Ullman, J.D., Harinarayan, V., Rajaraman, A.: Implementing data cubes efficiently. In: ACM SIGMOD International Conference on Management of Data, pp. 205–216 (1996)

[VS99] Vassiliadis, P., Sellis, T.: A survey of logical models for olap databases. SIGMOD Record 28, 65–69 (1999)

[VSM02] Verta, G., Stock, M., Morris, A.: Extending erd modeling notation to fuzzy management of gis data files. Data & Knowledge Engineering 40(2), 163–179 (2002)

Expressing Hierarchical Preferences in OLAP Queries

Panagiotis Chountas, Ermir Rogova, and Krassimir Atanassov

Abstract. OLAP query answering requirements for a knowledge based treatment of user requests led us to introduce the concept of hierarchical preferences over a universe that has a hierarchical structure. We introduce the automatic analysis of queries according to concepts defined as part of knowledge based hierarchy in order to guide the query answering as part of a data-warehouse environment with the aid of hierarchical Intuitionistic fuzzy sets, H-IFS. Based on the notion of H-IFS we propose an ad-hoc utility build on top of current OLAP tools like Oracle10g that allows us to enhance the query capabilities of by providing better and knowledgeable answers to user's requests. The theoretical aspects as well the practical issues and achieved results are presented throughout the rest of the paper.

1 Introduction

In 1970s, the need for flexible models and query languages to manage the ill-defined nature of information in Databases was identified [1]. Nowadays, the application of OLAP technology to other knowledge fields e.g., medical data and the use of semi-structured sources e.g., XML and non-structured sources has made these requirements even more important.

Over the past years we have witnessed an increasing interest in expressing user or domain preferences [2] inside database queries. First, it appeared to be desirable property of a query system to offer more expressive query languages that can be more faithful to what a user intends to say. Second, a classical query in the sense of relational paradigm may also have a restricted answer or sometimes an empty set of answers, while a relaxed version of the query enhanced with background or domain knowledge might be matched by some items in the database.

Frequently integrated DBMSs contain incomplete data which we may represent using hierarchical background knowledge to declare support contained in subsets

Panagiotis Chountas and Ermir Rogova
DKMG – HSCS, University of Westminster, Northwick Park, London, HA1 3TP, UK
e-mail: chountp@wmin.ac.uk

Krassimir Atanassov
CLBME, Bulgarian Academy of Sciences, Bl. 105, Sofia-1113, Bulgaria
e-mail: krat@argo.bas.bg

J. Kacprzyk et al. (Eds.): Uncer. Appro. for Spatial Data Model. and Process., SCI 271, 61–77.
springerlink.com

of the domain. These subsets may be represented in the database as partial values, which are derived from background knowledge using conceptual modelling to re-engineer the integrated DBMS.

Concerning query enlargement, several works such as [3] use a lattice of concepts to generalize unsolvable queries. An extended relational model for assigning possible values to an attribute value has been proposed by [4]. This approach may be used either to answer queries for decision making or for the extraction of answers and knowledge from relational databases. It is therefore important that appropriate functionality is provided for database systems to handle such information.

In studies about possibilistic ontologies [5], each term of an ontology is considered as a linguistic label and has an associated fuzzy description. Fuzzy pattern matching between different ontologies is then computed using these fuzzy descriptions. Studies about fuzzy thesauri have discussed different natures of relations between concepts. Fuzzy thesauri have been considered, for instance, in [6].

Recently in OLAP systems a need has been identified for enhancing the query scope with the aid of kind of relation that describe knowledge as well as ordering of the elements of a domain or a hierarchical universe.

However, in our context, the terms of the hierarchy [7], [8] [9] and the relations between terms are not fuzzy. These observations led us to introduce the concept of closure of the H-IFS which is a developed form defined on the whole hierarchy. The definition domains of the Hierarchical Intuitionistic Fuzzy sets, H-IFS that we propose below are subsets of hierarchies composed of elements partially ordered by the "kind of" relation or by $\subset$.

Based on the above observations, in this research, we particularly focus on incorporating hierarchical preferences expressed in the form of background-domain knowledge with the aim on enhancing the query scope and in return to get richer answer, closer to user requests.

We developed an ad-hoc OLAP utility known as 'IF-Oracle' [10] implemented on top of Oracle10g that allow us firstly to define and secondly incorporate hierarchical knowledge in the form of H-IFS as part of the standard SQL queries. We demonstrate the benefits of the 'IF-Oracle' by comparing the respective enhanced query answers against the Oracle10g standard query answers.

The rest of the paper is organised as follows; In section II we review the issue of value imprecision in OLAP Systems and emerging research challenges. In Section III we define the basic properties of Intuitionistic Fuzzy sets and H-IFS. In Section IV we define the extended SQL aggregators. In Section V we present and discussed the main concepts involved in the designing and implementation the 'IF-Oracle' ad-hoc utility and also demonstrate the potential of 'IF-Oracle' utility when it comes to query answering that requires utilisation of the domain knowledge in order to receive answer close to the user's intent. Finally we point to future research aims.

2 Value Imprecision in OLAP Systems

The need for flexible systems to manage value imprecision has been the focus for database researchers mainly in the context of the relational model. OLAP technology required [11] the extension of the relational systems with the inclusion of the data-cube and operators to operate over it. Alternatively, new models [12] were proposed to support OLAP based querying on top of multidimensional views. Both approaches support the organisation of data around several axes of analysis. In OLAP based systems, when it comes to the model level, support for value uncertainty will be required at the fact level as well at the level of dimensions with the support of non-rigid hierarchies [13]. Still [14] considers that facts and dimensions as in [15], [16] represent structural information.

Current research issues for OLAP systems can be summarised as follows:

i) Flexible models are required to support value uncertainty at fact level as well as at the dimension level with the provision of non rigid dimensions
ii) Flexibility should not be eliminated at the structural level. It should be allowed also at the query level. Users should be allowed to synthesise their own model of dimensions for analysis purposes based on existing structure. Dimensions may be based in either rigid or non-rigid hierarchies.

To this extent concepts can be used to describe how the data is organized in the data sources and to map such data to the concepts described in the Domain Ontology. These definitions are used to apply more extensively the business semantics described in the Domain Ontology, to support the rewrite of queries' conditions and combine OLAP features in this process. These semantics support the automatic recommendation of analysis according to the context of users' explorations in order to guide query answering, feature inexistent in current analytical tools.

Concepts are used to describe how the data is organized in the data sources and to map such data to the concepts described in the Domain Ontology. These definitions are used to apply more extensively the business semantics described in the Domain Ontology, to support the rewrite of queries' conditions and combine OLAP features in this process. These semantics support the automatic recommendation of analysis according to the context of users' explorations in order to guide query answering, feature inexistent in current analytical tools.

Realising a flexible OLAP environment where value uncertainty is accommodated at the level of models, give users much more flexibility when queries are imposed and at the same time expands the range of answers obtained in respond to those queries.

The main issues to be resolved are:

i) Imprecision at the level of multidimensional models: the semantics of value uncertainty need to been defined with regard to the main structures of multidimensional modelling (dimensions, hierarchies, facts) and the interrelationships between them. Also users or different applications should be able to define their own axes of analysis.
ii) Flexible Non-Deterministic Query System: this will allow the querying at the fact level with the assistance of OLAP operators after being re-defined with the aid of Intuitionistic fuzzy logic. More specifically we introduce the automatic analysis of queries according to concepts defined as part of a knowledge based hierarchy in order to guide the query answering as part of an integrated database environment with the aid of hierarchical Intuitionistic fuzzy sets, H-IFS.

Overall no significant attempt [17], [18] has been made for a generic representation of value uncertainty mainly as a property of axes of analysis and also as part of dynamic environment, where potential users may wish to define their "own" axes of analysis for querying either precise or imprecise facts. To put it differently, different users may wish to define their own dimensions of analysis based on a multidimensional model [19], [20]. In such cases, measured values and facts are characterised by descriptive values drawn from a number of dimensions, whereas values of a dimension are organised in a containment type hierarchy. This need is more obvious since we move from the classical DBMS environments to multi-source integrated environments where OLAP is the main query answering system.

We propose a unique ontological approach for the treatment of value uncertainty, with respect to multidimensional-OLAP modelling and flexible structuring of user defined versions of measures based on rigid or non-rigid-hierarchies.

3 Concept Based Hierarchies-Notion of H-IFS

Let us consider a sample concept based hierarchy named as wine. Temporarily let us ignore the meaning of weights.

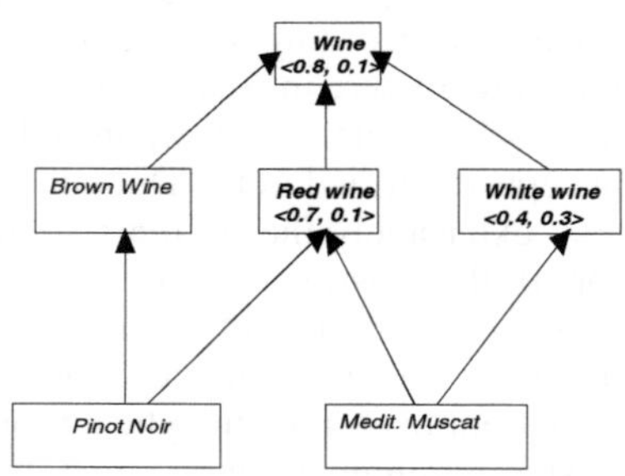

Fig. 1. Wine Hierarchy.

With respect to the Concept based hierarchy Wine, we try to express different ontological semantics, or "kind of" relations such as to what extent:

- Medit. Muscat is a "kind-of" Red wine?
- Medit. Muscat is a "kind-of" White wine?
- Pinot Noir is a "kind-of" Red wine?
- Pinot Noir is a "kind-of" Brown wine?
- Brown wine is a "kind-of" wine? Etc.

It is obvious from the above examples that if we wish to summarise the sales, for example, of products of "Brown wine" we need to take into account as well the fact that "Pinot Noir" may also be treated somehow as "Brown wine" when applying i.e. the SUM aggregator.

The above queries led us introduce the concept of closure of an Intuitionistic fuzzy set over a universe that has a hierarchical structure, which is a developed form defined on the whole hierarchy. For instance, in a query, if the user is interested in the element Wine, we consider that all kinds of Wine, Red wine, Brown wine, etc. are of interest. On the opposite, we consider that the super-elements (more general elements) of Wine in the hierarchy are too general to be relevant for the user's query.

3.1 *IFS- Notion of H-IFS*

Each element of an Intuitionistic fuzzy [21, 22] set has degrees of membership or truth (μ) and non-membership or falsity (ν), which don't sum up to 1.0 thus leaving a degree of hesitation margin (π).

As opposed to the classical definition of a fuzzy set given by A′ = $\tilde{A}=\{<x,\mu_A(x)>| x\in X\}$ where $\mu_A(x)\in[0,1]$ is the membership function of the fuzzy set A′, an Intuitionistic fuzzy set A is given by

$$A=\{<x,\mu_A(x),\nu_A(x)>| x\in X\}$$

$$\mu_A : X \to [0;1] \text{ and } \nu_A : X \to [0;1]$$

such that $0\le\mu_A(x)+\nu_A(x)\le 1$ and $\mu_A : X \to [0;1]$, $\nu_A : X \to [0;1]$ denote a degree of membership and a degree of non-membership of x ε A, respectively. Obviously, each fuzzy set may be represented by the following Intuitionistic fuzzy set

$A=\{<x,\mu_A(x),1-\mu_A(x)>| x\in X\}$For each Intuitionistic fuzzy set in X, we will call $\pi_A(x)=1-\mu_A(x)-\nu_A(x)$ an Intuitionistic fuzzy index (or a hesitation margin) of x ε A which expresses a lack of knowledge of whether x belongs to A or not. For each $x\in A$ $0\le\pi_A(x)\le 1$

Definition 1. Let A and B be two Intuitionistic fuzzy sets defined on a domain X. A is included in B (denoted A ⊆ B) if and only if their membership functions and non-membership functions satisfy the condition:

$$(\forall x\in X)\ (\mu_A(x)\le\mu_B(x)\ \&\ \nu_A(x)\ge\nu_B(x))$$

Two scalar measures are classically used in classical fuzzy pattern matching to evaluate the compatibility between an ill-known datum and a flexible query, known as

- a possibility degree of matching, **Π(Q/ D)**
- a necessity degree of matching, **N(Q/ D)**

Definition 2. Let Q and D be two Intuitionistic fuzzy sets defined on a domain X and representing, respectively, a flexible query and an ill-known datum:

The possibility degree of matching between Q and D, denoted Π(Q/D), is an "optimistic" degree of overlapping that measures the maximum compatibility between Q and D, and is defined by:

$$\Pi(Q/D) = \left\langle \sup_{x\in X} \min(1-\nu_Q(x), \nu_Q(x)), \inf_{x\in X} \max(1-\nu_D(x), \nu_D(x)) \right\rangle$$

The necessity degree of matching between Q and D, denoted N(Q/D), is a "pessimistic" degree of inclusion that estimates the extent to which it is certain that D is compatible with Q, and is defined by:

$$N(Q/D) = \left\langle \inf_{x\in X} \max(\mu_Q(x), 1-\mu_Q(x)), \sup_{x\in X} \min(\mu_D(x), 1-\mu_D(x)) \right\rangle$$

The problem occurring from defining Intuitionistic fuzzy sets based on the kind of relation is that two different Intuitionistic fuzzy sets on the same hierarchy do not necessarily have the same definition domain, which means they cannot be compared using the classic comparison operations Π(Q/ D), N(Q/ D)

3.1.1 The Notion of H-IFS

The definition domains of the hierarchical fuzzy sets [23, 24, 25] that we propose below are subsets of hierarchies composed of elements partially ordered by the "kind of" relation. An element l_i is more general than an element l_j (denoted $l_i \sim l_j$), if l_i is a predecessor of l_j in the partial order induced by the "kind of" relation of the hierarchy. An example of such a hierarchy is given in Fig..2. A hierarchical Intuitionistic fuzzy set is then defined as follows.

Definition 3. Let F be a H-IFS defined on a subset D of the elements of a hierarchy L. It degree is denoted as <μ, ν>. The closure of F, denoted clos(F), is a H-IFS defined on the whole set of elements of L and its degree $<\mu, \nu>_{clos(F)}$ is defined as follows.

For each element **l** of L, let $S_L = \{l_1, \ldots, l_n\}$ be the set of the smallest super-elements in D.

If S_L is not empty,

$$<\mu,\nu>_{clos(F)} (S_L) = < \max_{1\le i\le n} \mu(L_i), \min_{1\le i\le n} \nu(L_i) >$$

else

$$<\mu,\nu>_{clos(F)} (S_L) = <0,0>$$

In other words, the closure of a H-IFS F is built according to the following rules. For each element l_1 of L:

- If l_I belongs to F, then l_I keeps the same degree in the closure of F (case where $S_L = \{ l_I \}$).
- If l_I has a unique smallest super-element l_I in F, then the degree associated with l_I is propagated to L in the closure of F, $S_L = \{ l_I \}$ with $l_I > l_I$)

If L has several smallest super-elements $\{l_1, \ldots, l_n\}$ in F, with different degrees, a choice has to be made concerning the degree that will be associated with l_I in the closure. The proposition put forward in definition 3, consists of choosing the maximum degree of validity μ and minimum degree of non validity ν associated with $\{l_1, \ldots, l_n\}$. We refer to as the *Optimistic strategy*.

We can also utilise a *Pessimistic strategy* which consists of choosing the minimum degree of validity μ and maximum degree of non validity ν associated with $\{l_1, \ldots, l_n\}$.Alternatively, an *Average strategy* could be utilised, which consists of calculating the IF-Average and applying it to the degrees of validity μ and non-validity ν.

It has been observed that two different H-IFSs, defined on the same hierarchy, can have the same closure, as in the following example.

The H-IFSs Q={Wine<1,0>, Red Wine<0.7,0.1>, Brown Wine<1,0>, White Wine <0.4,0.3>} and

R ={Wine<1,0>, Red Wine<0.7,0.1>, Brown Wine<1,0>, Pinot Noir <0.4,0.3>} have the same closure, represented Fig.2 below.

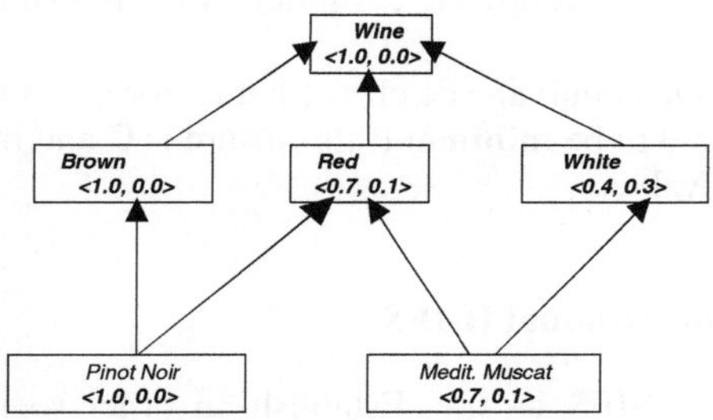

Fig. 2. Common closure of the H-IFS's Q and R

Such H-IFSs form equivalence classes with respect to their closures.

Definition 4. Two H-IFSs Q and R, defined on the same hierarchy, are said to be equivalent Q≡R if and only if they have the same closure.

Property. Let Q and R be two equivalent Intuitionistic hierarchical fuzzy sets. If $l_I \in \text{dom}(Q) \cap \text{dom}(R)$, then $<\mu,\nu>(Q.l_I) = <\mu,\nu>(R.l_I)$

Proof. According to the definition of the closure of a H-IFS F, *definition 3*, the closure of F preserves the degrees that are specified in F. As Q and R have the same closure (by definition of the equivalence), an element that belongs to Q and R necessarily has the same degree <μ,ν> in both.

We can note that R contains the same element as Q with the same <μ,ν>, and also one more element Pinot Noir<1,0>. The <μ,ν> associated with this additional element is the same as in the closure of Q. Then it can be said that the element, Pinot Noir<1,0> is derivable in R through Q. The same conclusions can be drawn in the case of Medit. Muscat <0.7, 0.1>

Definition 5. Let F be a hierarchical fuzzy set, with dom(F) = $\{l_1, \ldots, l_n\}$, and F_{-k} the H-IFS resulting from the restriction of F to the domain dom(F) \ $\{l_k\}$. l_k is deducible in F if

$$<\mu, \nu>clos_{(F-k)}(l_k) = <\mu, \nu>clos_{(F)}(l_k)$$

As a first intuition, it can be said that removing a derivable element from a hierarchical fuzzy set allows one to eliminate redundant information. But, an element being derivable in F does not necessarily mean that removing it from F will have no consequence on the closure: removing k from F will not impact the degree associated with k itself in the closure, but it may impact the degrees of the sub-elements of k in the closure.

For instance, if the element Brown Wine is derivable in Q, according to *definition 5*, removing Brown Wine <1,0> from Q would not modify the degree of Brown Wine itself in the resulting closure, but it could modify the degree of its sub-element Pinot Noir. Thus, Brown Wine <1,0> cannot be derived or removed. This remark leads us to the following definition of a minimal hierarchical fuzzy set.

Definition 6. In a given equivalence class (that is, for a given closure C), a hierarchical fuzzy set is said to be **minimal** if its closure is C and if none of the elements of its domain is derivable.

3.1.2 Obtaining the Minimal H-IFS

Step 1: Assign Min-H-IFS $\leftarrow \emptyset$. Establish an order so that the sub-elements $\{l_1,\ldots,l_n\}$ of the hierarchy L are examined after its super-elements.

Step 2: Let l_1 be the first element and $(\mathbf{l_1})/<\mu, \nu> \neq (\mathbf{l_1})/<\mathbf{0}, \mathbf{0}>$ then add $\mathbf{l_1}$ to Min-H-IFS and $<\mu, \nu>_{\mathbf{clos(Min\text{-}HIFS)}} (\mathbf{l_1}) = (\mathbf{l_1})/<\mu, \nu>$.

Step 3: Let us assume that K elements of the hierarchy L satisfy the condition $<\mu, \nu>_{\mathbf{clos(Min\text{-}HIFS)}} (\mathbf{l_i}) = (\mathbf{l_i})/<\mu, \nu>$. In this case the Min-H-IFS do not change. Otherwise go to next element $\mathbf{l_{k+1}}$ and execute Step 4.

Step 4: The $\mathbf{l_{k+1}}/<\mu_{k+1}, \nu_{k+1}>$ associated with $\mathbf{l_{k+1}}$. In this case $\mathbf{l_{k+1}}$ is added to Min-H-IFS with the corresponding $<\mu_{k+1}, \nu_{k+1}>$.

Step 5: Repeat steps three and four until $\mathbf{clos_{(Min\text{-}HIFS)}}=\mathbf{C}$.

For instance the H-IFSs S_1 and S_2 are **minimal** (none of their elements is derivable). They cannot be reduced further.

S_1= Wine<1,0>
S_2={Wine<1,0>, Red Wine<0.7,0.1>, Pinot Noir<1,0>, White Wine <0.4, 0.3>}

3.1.3 Representing H-IFS as Concept Relations

The structure of any H-IFS can be described by a domain concept relation DCR = (Concept, Element), where each tuple describes a relation between elements of the domain on different levels. The DCR can be used in calculating recursively [26] the different summarisation or selection paths as follows:

$$PATH \leftarrow DCR_{\{x=1...(n-2) \mid n>2\}} \bowtie DCR_x$$

If $n \leq 2$, then DCR becomes the Path table as it describes all summarisation and selection paths. These are entries to a knowledge table that holds the metadata on parent-child relationships. An example is presented below

Table 1. Domain Concept Relation

DCR	
Concept	**Element**
Wine <1.0, 0.0>	Brown Wine <1.0, 0.0>
Wine <1.0, 0.0>	Red Wine <0.7, 0.1>
Wine <1.0, 0.0>	White Wine <0.4, 0.3>
Brown Wine <1.0, 0.0>	Pinot Noir <1.0, 0.0>
Red Wine <0.7, 0.1>	Pinot Noir <1.0, 0.0>
Red Wine <0.7, 0.1>	Medit. Muscat <0.7, 0.1>
White Wine <0.4, 0.3>	Medit. Muscat <0.7, 0.1>

Table 1 shows how our Wine hierarchy knowledge table is kept. Paths are created by running a recursive query that reflects the 'PATH' algebraic statement. The hierarchical IFS used as example throughout this paper comprises of 3 levels, thus calling for the SQL-like query as below:

SELECT A.Concept as Grand-concept, b.concept, b.element
FROM DCR as A, DCR as B
WHERE A.child=B.parent;

This query will produce the following paths:

Table 2. Path Table

Path			
Grand-concept	**Concept**	**Element**	**Path Colour**
Wine <1.0, 0.0>	Brown Wine <1.0, 0.0>	Pinot Noir <1.0, 0.0>	Red
Wine <1.0, 0.0>	Red Wine <0.7, 0.1>	Pinot Noir <1.0, 0.0>	Blue
Wine <1.0, 0.0>	Red Wine <0.7, 0.1>	Medit. Muscat <0.7, 0.1>	Green
Wine <1.0, 0.0>	White Wine <1.0, 0.0>	Medit. Muscat <0.7, 0.1>	Brown

Table 2 presents a pictorial view of the four distinct summarisation and selection paths. These paths will be used in fuzzy queries to extract answers that could be either definite or possible. This will be realised with the aid of the predicate (θ).A predicate (θ) involves a set of atomic predicates ($\theta_1, \ldots, \theta_n$) associated with the aid of logical operators p (i.e. $\wedge$, $\vee$, etc.). Consider a predicate θ that takes the value "Red Wine", θ = "Red Wine".

After utilizing the IFS hierarchy presented in Fig.2, this predicate can be reconstructed as follows:

$$\theta = \theta_1 \vee \theta_2 \vee \ldots \vee \theta_n$$

In our example, θ_1="Red Wine", θ_2="Pinot Noir" and θ_n="Medit. Muscat". The reconstructed predicate θ = (Red Wine $\vee$ Pinot Noir $\vee$ Medit. Muscat) allows the query mechanism to not only definite answers, but also possible answers [27].

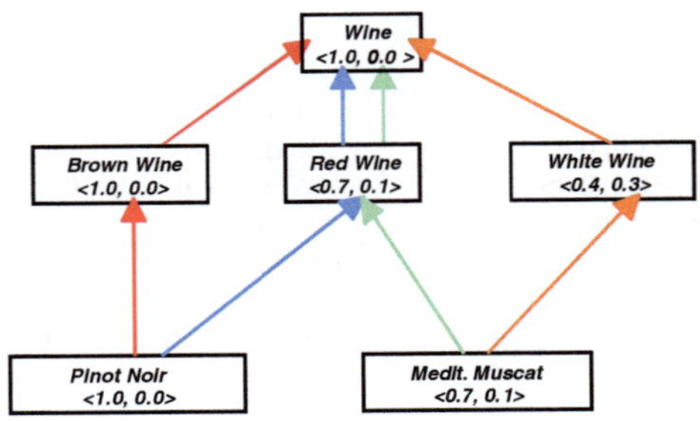

Fig. 3. Pictorial representation of paths

In terms a query retrieving data from a summary table, the output contains not only records that match the initial condition, but also those that satisfy the reconstructed predicate. Consider the case where no records satisfy the initial condition (Red Wine). Traditional aggregation query would have returned no answer, however, based on our approach, the extended query would even in this case, return an answer, though only a possible one, with a specific belief and disbelief $<\mu, \nu>$. It will point to those records that satisfy the reconstructed predicate θ, more specifically, "Pinot Noir and Medit. Muscat".

Following the representation of H-IFS as concept relations and the definition of summarisation paths, there is still a need to extend the traditional aggregation operators in order to cope with flexible hierarchies of data organisations.

4 Extended Relational Aggregation Operators

Aggregation ***(A)*:** An aggregation operator A is a function $A(G)$ where $G = \{<x, \mu_F(x), \nu_F(x)> \mid x \in X\}$ where $x=<att_1, \ldots, att_n>$ is an ordered tuple belonging to a given universe X, $\{att_1, \ldots, att_n\}$ is the set of attributes of the elements of X, $\mu_F(x)$ *and* $\nu_F(x)$ are the degree of membership and non-membership of x. The result is a bag of the type $\{<x', \mu_F(x'), \nu_F(x')> \mid x' \in X\}$. To this extent, the bag is a group of elements that can be duplicated and each one has a degree of μ and ν.

- *Input:* $R_i = (l, F, H)$ and the function $A(G)$
- *Output:* $R_o = (l_o, F_o, H_o)$ *where*

- l is a set of levels $l_1,\ldots, l_n$, that belong to a partial order $\leq O$ To identify the level l as part of a hierarchy we use *dl.*
- $l_\bot$: base level, $l_\top$: top level

For each pair of levels l_i and l_j we have the relation

$$\mu_{ij} : l_i \times l_j \rightarrow [0,1], \quad \nu_{ij} : l_i \times l_j \rightarrow [0,1,] \quad 0 < \mu_{ij} + \nu_{ij} < 1$$

- F is a set of fact instances with schema $F = \{<x, \mu_F(x), \nu_F(x)> \mid x \in X\}$, where $x=<att_1, \ldots, att_n>$ is an ordered tuple belonging to a given universe X, $\mu_F(x)$ *and* $\nu_F(x)$ are the degree of membership and non-membership of x in the fact table F respectively.
- H is an object type history that corresponds to a structure(l, F, H') which allows us to trace back the evolution of a structure after performing a set of operators i.e. aggregation

The definition of the extended group operators allows us to define the extended group operators ***Roll up (Δ), and Roll Down (Ω).***

Roll up (Δ): The result of applying Roll up over dimension d_i at level dl_r using the aggregation operator A over a relation $R_i=(l_i, F_i, H_i)$ is another relation $R_o=(l_o, F_o, H_o)$

Input: $R_i = (l_i, F_i, H_i)$
Output: $R_o = (l_o, F_o, H_o)$

An object of type history is a recursive structure:

$$\begin{cases} \omega \text{ is the initial state of the relation.} \\ (l, A, H') \text{ is the state of the relation after performing an operation on it.} \end{cases}$$

The structured history of the relation allows us to keep all the information when applying *Roll up* and get it all back when *Roll Down* is performed. To be able to apply the operation of *Roll Up* we need to make use of the IF_{SUM} aggregation operator.

Roll Down (Ω): This operator performs the opposite function of the *Roll Up* operator. It is used to roll down from the higher levels of the hierarchy with a greater degree of generalization, to the leaves with the greater degree of precision. The result of applying *Roll Down* over a relation $R_i = (l, F, H)$ having $H=(l', A', H')$ is another relation $R_o= (l', F', H')$.

Input: $R_i=(l, F, H)$
Output: $R_o=(l', F', H')$ where *F' → set of fact instances defined by operator A.*

To this extent, the *Roll Down* operative makes use of the recursive history structure previously created after performing the *Roll Up* operator.

The definition of aggregation operator points to the need of defining the IF extensions for traditional group operators [12], such as *SUM, AVG and MAX.* Based on the standard group operators, we provide their IF extensions and meaning.

IF_{SUM}: The IF_{sum} aggregate, like its standard counterpart, is only defined for numeric domains. The relation *R* consists of tuples R_i *with* $1 \le i \le m$. The tuples R_i are assumed to take Intuitionistic Fuzzy values for the attribute att_{n-1} for $i = 1$ *to* m we have $R_i[att_{n-1}] = \{<\mu_i(u_{ki}), \nu_i(u_{ki})>/ u_{ki} \mid 1 \le k_i \le n\}$. The IF_{sum} of the attribute att_{n-1} of the relation R is defined by:

$$IFS_{SUM}((att_{n-1})(F)) = \left\{ <u>/y \mid ((u = \min_{i=1}^{m}(\mu_i(u_{ki}), \nu_i(u_{ki})) \wedge \left(y = \sum_{k_i=k_1}^{km} u_{ki} \right) (\forall k_1,\ldots,k_m : 1 \le k_1,\ldots,k_m \le n)) \right\}$$

IF_{AVG}: The IF_{AVG} aggregate, like its standard counterpart, is only defined for numeric domains. This aggregate makes use of the IF_{SUM} that was discussed previously and the standard *COUNT.* The IF_{AVG} can be defined as:

$$IF_{AVG}((att_{n-1})(R) =$$

$$IF_{SUM}((att_{n-1})(R)) \,/\, COUNT((att_{n-1})(R))$$

IFS_{MAX}: The IFS_{MAX} aggregate, like its standard counterpart, is only defined for numeric domains. Given a fact F defined on the schema X $(att_1, \ldots, att_n)$, let att_{n-1} defined on the domain $U=\{u_1, \ldots, u_n)$. The fact *F* consists of fact instances f_i *with* $1 \le i \le m$.

The fact instances f_i are assumed to take Intuitionistic Fuzzy values for the attribute att_{n-1} for $i = 1$ *to* m we have $f_i[att_{n-1}] = \{<\mu_i(u_{ki}), \nu_i(u_{ki})>/ u_{ki} \mid 1 \le k_i \le n\}$. The IFS_{MAX} of the attribute att_{n-1} of the fact table F is defined by:

$$IFS_{MAX}((att_{n-1})(F)) = \{<u>/y \mid ((u = \min_{i=1}^{m}(\mu_i(u_{ki}), \nu_i(u_{ki})) \wedge (y = \max_{i=1}^{m}(\mu_i(u_{ki}), \nu_i(u_{ki}))) (\forall k_1,\ldots,k_m : 1 \le k_1,\ldots,k_m \le n))\}$$

In the next section we demonstrate the usefulness of the H-IFS notion and the extended aggregation operators for extending the query capabilities of Oracle10g. We developed an ad-hoc utility 'IF-Oracle' implemented on top of Oracle10g that allow us to:

- Define an H-IFS hierarchy
- Incorporate hierarchical knowledge in the form of H-IFS as part of the standard SQL queries.
- Enhance the scope of query answers against the Oracle10g standard query answers.

5 IF-Oracle Ad Hoc Utility

IF-Oracle has been developed using Visual Studio.Net as an ad-hoc utility that is attached to and enhances Oracle10g DBMS query capabilities. For demonstrating the functionality of IF-Oracle let us consider a sample multidimensional model, Fig.3 in the form of a star schema that describes sales of Vitis Vinifera type wines.

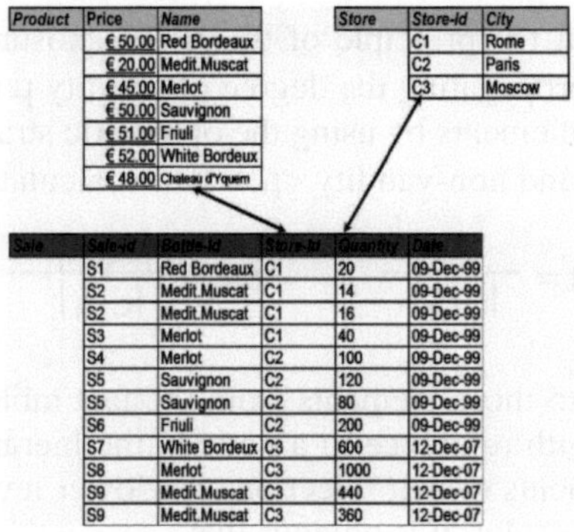

Product	Price	Name
	€ 50.00	Red Bordeaux
	€ 20.00	Medit.Muscat
	€ 45.00	Merlot
	€ 50.00	Sauvignon
	€ 51.00	Friuli
	€ 52.00	White Bordeux
	€ 48.00	Chateau d'Yquem

Store	Store-Id	City
	C1	Rome
	C2	Paris
	C3	Moscow

Sale	Sale-Id	Bottle-Id	Store-Id	Quantity	Date
	S1	Red Bordeaux	C1	20	09-Dec-99
	S2	Medit.Muscat	C1	14	09-Dec-99
	S2	Medit.Muscat	C1	16	09-Dec-99
	S3	Merlot	C1	40	09-Dec-99
	S4	Merlot	C2	100	09-Dec-99
	S5	Sauvignon	C2	120	09-Dec-99
	S5	Sauvignon	C2	80	09-Dec-99
	S6	Friuli	C2	200	09-Dec-99
	S7	White Bordeux	C3	600	12-Dec-07
	S8	Merlot	C3	1000	12-Dec-07
	S9	Medit.Muscat	C3	440	12-Dec-07
	S9	Medit.Muscat	C3	360	12-Dec-07

Fig. 3. Sample of a Star Schema.

Fig.4 shows a sub-hierarchy that has been derived from the Vitis Vinifera domain for testing purposes. On the left it is shown the tree structure view as displayed in IF-Oracle, while on the right we have shown the tree representation.

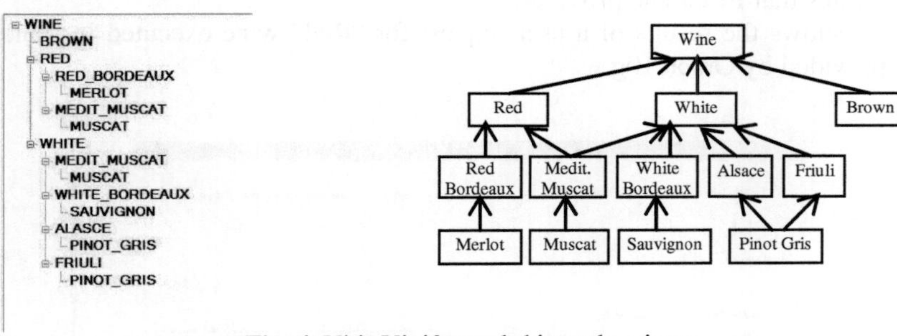

Fig. 4. Vitis Vinifera sub-hierarchy views.

After forming the structure and storing it as a concept relation in Oracle10g, we perform the calculation of the hierarchical closure of the H-IFS and its weights.

The user now has the choice of selecting three different strategies: ***Optimistic, Pessimistic*** or ***Average*** as defined on section 3.1.1.

Let's assume that the user's interest lays on finding information about Red wines.

Fig.5 below shows the hierarchy after weights have been calculated and assigned reflecting the user's intent.

Fig. 5. Vitis Vinifera sub-hierarchy view with weights.

We can observe that the principle of the H-IFS closure *(see definition 3)* has been preserved when propagating the degree of validity μ and non-validity ν from super-elements to sub-elements by using the optimistic strategy.
The degree of validity and non-validity <μ, ν > are calculated as follows:

$$\mu = \frac{|c_l|}{|c_{l-1}|} \qquad \nu = \frac{|\neg c_l|}{|c_{l-1}|}$$

Where c_l corresponds to those elements from the fact table that absolutely satisfy the selection criteria with reference to a node in the hierarchy. C_{l-1} represents the elements children elements of that selection on a lower level that satisfy the selection condition to some extent. It is obvious that

$$\pi = 1 - (\mu + \nu)$$

After adding the hierarchy into the repository and automatically calculating the weights for the requested nodes, the user can utilize the ad-hoc interface for execution of queries either in standard SQL or make use of the enhanced Select clause and features that IF-Oracle provides.

Fig.6 shows the results of a user request for "Red" wine executed in standard SQL provided by Oracle10g.

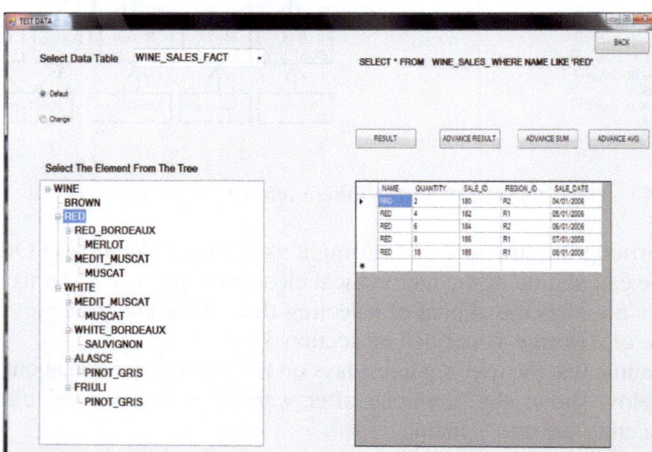

Fig. 6. Standard SQL output for "Red" wine

In contrast, Fig.7 shows the output after executing the same query, but this time using the IF-Oracle utility.

By comparing the two figures, one can observe that IF-Oracle produces a knowledge-based answer instead of mindlessly matching the records against the word "Red".

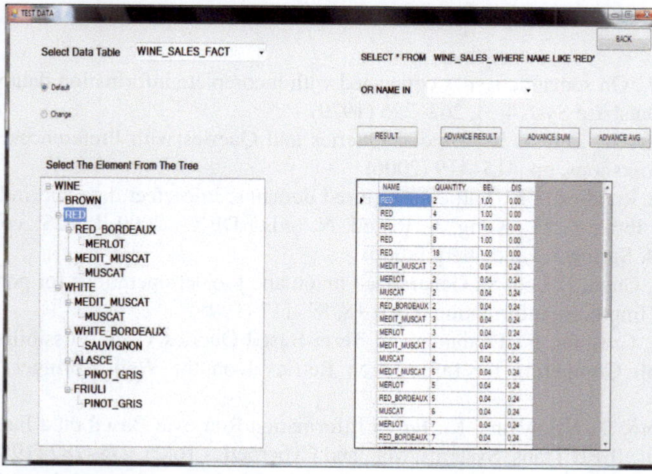

Fig. 7. Enhanced SQL output for "Red" wine

The results show that IF-Oracle not only retrieves sales of "Red" bottles, but also sales of bottles that are classified as red wines by the knowledge represented in the H-IFS hierarchy as "Merlot", "Red Bordeaux", "Medit. Muscat", etc. with indicative degrees of $<\mu, \nu>$ relevant to the user's preference.

At this point one may decide to further enhance the query capabilities of the IF-Oracle utility by allowing versions of hierarchies. In such case similarities [28] and dissimilarities [29] between different versions should be reflected in the query results.

6 Conclusions

In this paper, we focus on integrating hierarchical preferences in OLAP queries expressed in the form of background-domain knowledge with the aim on enhancing the query scope and in return to get richer answer, closer to user requests. We provide a means of using background knowledge to re-engineer query processing and answering with the aid of H-IFS and Intuitionistic Fuzzy relational representation.

The hierarchical links defined on the basis of the H-IFS closure are representing knowledge in different forms. The membership of an element in a H-IFS has consequences on the membership and non-membership of its sub elements in this set.

We demonstrated the simplicity and implement-ability of the H-IFS notion by adding an ad-hoc utility 'IF-Oracle' in Oracle10g that allow us to enrich the scope of query and receive answers closer to user's intent and preferences even when answers are not obvious when using the standard SQL provided by Oracle10g.

Future research efforts will concentrate on incorporating knowledge arriving from external sources either semi structured or unstructured i.e. WordNet or Wikipedia considering the web as such as source. Furthermore considering the notion of the minimal H-IFS one could consider to devise new optimisation techniques for making query processing more efficient.

References

1. Lipski, J.: On semantic issues connected with incomplete information databases. ACM Trans. Database Syst. 4(3), 262–296 (1979)
2. Slawomir, Z., Janusz, K.: Bipolar Queries and Queries with Preferences. In: DEXA 2006 Workshops, pp. 415–419 (2006)
3. Rice, S., Roddick, J.F.: Lattice-structured domains, imperfect data and inductive queries. In: Ibrahim, M., Küng, J., Revell, N. (eds.) DEXA 2000. LNCS, vol. 1873, pp. 664–674. Springer, Heidelberg (2000)
4. Bell, D., Guan, J., Lee, S.: Generalized union and project operations for pooling uncertain and imprecise information. DKE 18, 89–117 (1996)
5. Pasi, G., Crestani, F.: Evaluation of Term-Based Queries Using Possibilistic Ontologies. Soft Computing for Information Retrieval on the Web. Springer, Heidelberg (2005)
6. Miyamoto, S., Nakayama, K.: Fuzzy Information Retrieval Based on a Fuzzy Pseudothesaurus. IEEE Trans. Systems, Man and Cybernetics 16(2), 278–282 (1986)
7. Rogova, E., Chountas, P., Atanassov, K.: Flexible hierarchies and fuzzy knowledge-based OLAP. In: FSKD 2007, vol. 2, pp. 7–11. IEEE Computer Society Press, Los Alamitos (2007)
8. Chountas, P.: On Intuitionistic fuzzy sets over universes with hierarchical structures. Notes on Intuitionistic Fuzzy Sets 13(1), 52–56 (2007)
9. Rogova, E., Chountas, P.: On imprecision Intuitionistic fuzzy sets & OLAP – The case for KNOLAP. In: IFSA 2007, pp. 11–23. Springer, Heidelberg (2007)
10. Chountas, P., Rogova, E., Atanassov, K., Mohammed, S.: The Notion of H-IFS -An Approach for Enhancing Query Capabilities in Oracle10g. In: IEEE Intelligent Systems 2008, pp.13-8–13-13. IEEE Computer Society Press, Los Alamitos (2008)
11. Pedersen, T.B., Jensen, C.S.: Multidimensional Data Modeling for Complex Data. In: Proc. of 15th ICDE, pp. 336–345. IEEE Computer Society, Los Alamitos (1999)
12. Gray, J., Bosworth, A., Layman, A., Pirahesh, H.: Data cube: a relational aggregation operator generalizing group-by, cross-tabs and subtotals. Journal of Data Mining and Knowledge Discovery 1(1), 29–53 (1997)
13. Delgado, M., Molina, C., Sanchez, D., Vila, A., Rodriguez-Ariza, L.: A fuzzy multidimensional model for supporting imprecision in OLAP. In: Proc. of FUZZ-IEEE, vol. 3, pp. 1331–1336 (2004)
14. Chamoni, P., Stock, S.: Temporal Structures in Data Warehousing. In: Mohania, M., Tjoa, A.M. (eds.) DaWaK 1999. LNCS, vol. 1676, pp. 353–358. Springer, Heidelberg (1999)
15. Pedersen, T.B., Jensen, C.S., Dyreson, C.E.: A foundation for capturing and querying complex multidimensional data. Information Systems 26(5), 383–423 (2001)
16. Mendelzon, A., Vaisman, A.: Temporal Queries in OLAP. In: Proc. of the 26th VLDB 2000, ACM Portal, pp. 242–253 (2000)
17. Chountas, P., Petrounias, I., Vasilakis, C., Tseng, A., El-Darzi, E., Atanassov, K.T., Kodogiannis, V.S.: On Uncertainty and Data-Warehouse Design. In: Yakhno, T. (ed.) ADVIS 2004. LNCS, vol. 3261, pp. 4–13. Springer, Heidelberg (2004)
18. Eder, J., Koncilia, C.: Changes of Dimension Data in Temporal Data Warehouses. In: Kambayashi, Y., Winiwarter, W., Arikawa, M. (eds.) DaWaK 2001. LNCS, vol. 2114, pp. 284–293. Springer, Heidelberg (2001)
19. Kimball, R.: The Data Warehouse Toolkit. John Wiley & Sons, New York (1996)

20. Inmon, W.H.: Building the Data Warehouse, 2nd edn. John Wiley & Sons, New York (1996)
21. Atanassov, K.: Intuitionistic Fuzzy Sets. Springer, Heidelberg (1999)
22. Atanassov, K.: Remarks on the Intuitionistic fuzzy sets. Fuzzy Sets and Systems 51(1), 117–118 (1992)
23. Kolev, B., Chountas, P., Rogova, E., Atanassov, K.: Representation of Value Imperfection with the Aid of Background Knowledge. H-IFS Studies in Computational Intelligence, vol. 109, pp. 473–494. Springer, Heidelberg (2008)
24. Rogova, E., Chountas, P., Atanassov, K.: The Notion of H-IFS in Data Modelling. In: FUZZ-IEEE 2008, pp. 1397–1403. IEEE Computer Society Press, Los Alamitos (2008)
25. Chountas, P., Atanasov, K., Rogova, E.: H-IFS: Modelling & Querying over Hierarchical Universes. In: International Conference on Information Processing And Management of Uncertainty-IPMU 2008, pp. 1628–1634 (2008) ISBN 978-84-612-3061-7
26. Silberschatz, A., Korth, H., Sudarshan, S.: Database System Concepts. McGraw-Hill, New York (2006)
27. Rundensteiner, E., Bic, L.: Aggregates in posibilistic databases. In: VLDB 1989, pp. 287–295 (1989)
28. Szmidt, E., Kacprzyk, J.: Distances between Intuitionistic fuzzy sets. Fuzzy Sets and Systems 114(3), 505–518 (2000)
29. Szmidt, E., Kacprzyk, J.: A new concept of a similarity measure for intuitionistic fuzzy sets and its use in group decision making. In: Torra, V., Narukawa, Y., Miyamoto, S. (eds.) MDAI 2005. LNCS (LNAI), vol. 3558, pp. 272–282. Springer, Heidelberg (2005)

Imperfect Multisource Spatial Data Fusion Based on a Local Consensual Dynamics

Gloria Bordogna, Marco Pagani, and Gabriella Pasi

Abstract. Strategies for multisource spatial data fusion have generally to cope with distinct kinds of uncertainty, related to both the trust of the information source, the imperfection of spatial data, and the vagueness of the fusion strategy itself. In this chapter we propose a consensual fusion method that allows to flexibly model several fusion strategies ranging from a risk-taking to a risk-adverse attitude, and capable to cope with both data imprecision and source reliability. Uncertainty and imprecision in spatial data are represented by associating a fuzzy value with each spatial unit. The fusion function models a consensual dynamics and is parameterized so as to consider a varying spatial neighborhood of the data to fuse. Moreover the fusion has a quantifier-guided nature, reflecting the concept of a fuzzy majority and works on imprecise values to compute an imprecise result. It is formalized by a generalized OWA operator defined in the paper for aggregating imprecise values with distinct importance. The consensual fusion works so that the greater the trust score of the source and its agreement with the other sources, the more influent (important) is the data from the source in determining the consensual values. Thus the obtained fused map is determined in each location by a distinct majority of the sources, those that locally are in agreement. In cases where the data are affected by uncertainty one can require to fuse them so as to compute a result affected by at most a given maximum uncertainty level.

1 Introduction

Spatial data fusion consists of a data integration process that combines spatial data from multiple sources to generate spatial data of "higher quality", carrying information not available from any individual source [28]. "Higher quality" can be intended as either data that support a better description of a spatial feature, or a better signal, or even a better decision. This last interpretation is the one that we model in this chapter.

Gloria Bordogna and Marco Pagani
CNR IDPA, via Pasubio 5, 24044 Dalmine (BG) Italy

Gabriella Pasi
Università di Milano Bicocca, viale Sarca 336, 20126 Milano, Italy

J. Kacprzyk et al. (Eds.): Uncer. Appro. for Spatial Data Model. and Process., SCI 271, pp. 79–94.
springerlink.com

We consider the fusion of spatial data that are independently produced by either software models or human experts, hereafter named sources, for example to classify the territory into risk/hazard maps with the objective of achieving a more robust decision map that synthesizes the consensual opinion of the experts/models; note that, individual sources can produce conflicting maps. In the fusion, we want to take into account the distinct trusts or presumed credits of the sources, that can be stated by a decision maker responsible of defining the fusion strategy, the imperfection, i.e. imprecision and uncertainty, of the data, the vagueness of the fusion strategy itself, i.e., the decision maker's attitude, and the spatial consensus among the data.

Current GISs are inadequate to support the experts in modeling multisource spatial data fusion affected by uncertainty because flexible decision strategies can hardly be defined by using the available aggregation operators [7][18][20]. The fusion operations that GISs offer are generally based on Boolean logic, basically maps overlay and weighted linear combination [19]. Further, these systems do not represent and manage the imprecision and uncertainty of the data, allowing to associate only precise values with each spatial unit. These are the reasons that motivated our work whose final objective would be to develop a software module within a GIS allowing the user to define and then execute his/her personalized fusion strategy on available data, possibly affected by imperfection [4].

The definition of a spatial data fusion strategy first requires to represent the input data in a common space, and then to define the way in which these data must be combined to generate the output. The first step is necessary when the input data are heterogeneous, as in the case of multisource data, and characterized by either different resolution, measurement errors, range of values, or distinct reliability of their source [27]. These are all causes of imprecision and uncertainty that must be appropriately dealt with when fusing spatial information.

Distinct models of information fusion have been proposed. A rich survey of multi criteria fusion literature can be found in [1][9][19][27].

Fuzzy set theory was applied to information fusion to flexibly model the fusion criterion by means of aggregation operators that can be defined with varying trade-offs between a severe and an indulgent behavior [9][11][13][22][31]. Coupled with possibility theory, uncertain information can be represented and managed [10]. These approaches are appealing since they can be defined to model a variety of real situations in a flexible way [6][8][15].

In this context we formalize our proposal of a soft consensual fusion model that manages both the trust of the sources and the imperfection of data. The soft fusion strategy is based on a linguistic quantifier [33] associated with the concept of a fuzzy majority [16] and implemented by a generalization of the OWA operator [29][30]. OWA operators have been indicated as appropriate tools for spatial data fusion since they are a family of mean-like operators that allow implementing distinct fusion strategies [19][22].

The fusion has the objective of synthesizing the results independently produced by the sources, i.e., the experts or the models, by taking into account their agreements, i.e., the data values variability within a local or global spatial neighborhood, so as to reduce possible semantic errors. The fusion operator is then a context dependent operator according to the classification given in [1]. Moreover it takes into account the data sources trust scores, that is, their reliability or presumed credit. These characteristics are very important in the spatial context where data may come from distinct sources with very distinct reliability, and acquisition characteristics.

The approach is robust in the sense that it copes with data of different types possibly affected by imperfection, such as measurement errors of the means of acquisition, approximation due to the adopted representation, incompleteness, etc. To this end, data can be represented by fuzzy values interpreted as possibility distributions [5], and when fusing them, one can specify a maximum tolerable level u of uncertainty of the result. The uncertainty level u specifies the u-cut that must be applied to the fuzzy values so as to generate imprecise data values, i.e., intervals of basic values. The generalized OWA operator defined in this paper fuses these intervals into an imprecise result that is affected by the level of uncertainty u.

In this paper we consider the fusion of spatial data represented in raster form, i.e., grid data. In section II we introduce the type of imperfect data managed and its representation within the fuzzy set framework. In section III, the consensual fusion model is described and formalized. In section IVI, an application example to generate a consensual fused seismic map is discussed. Finally the conclusions summarize the main achievements and future perspectives.

2 Imperfection in Spatial Data

In this section we analyze spatial data with respect to both their "imperfection" intended as either imprecision or uncertainty [3][5], and their relationships relevant to the fusion problem [31]. As far as the pixel values are concerned, they can be specified with one among three types:

Numeric values: for this type of data a metrics and all types of arithmetic operations are defined. Examples are the local slope or altitude of a spatial position; the density of some spatial property such as population, pollution, etc.;

Ordinal values: for this type of data an order and composition operations on the index of the labels, ex. similarity or proximity relationship between indexes, are defined. Examples are the classes of hazard, risk, susceptivity etc.

Nominal values: for this type of data the composition of values is meaningless; nevertheless, a similarity or proximity relationship between each pairs of values can be defined to represent a physical/chemical property of the data. For example the names of soil types and lithology types can be ordered to reflect their favorability to contribute to the occurrence of landslides.

Imperfection in spatial data may affect either the pixel values or the pixels' spatial reference. In the following, we analyze the representation within fuzzy set theory and possibility theory of the different kinds of imperfection of the spatial data.

Imperfection of Numeric Values

Numeric values can be imprecise and uncertain when they are not single elements of the numeric reference domain. This is the case of values obtained by statistic analysis expressing mean values of a property with an associated dispersion, such as daily temperatures. Within fuzzy set theory, uncertain and imprecise numeric values are represented by fuzzy subsets on their basic numeric domains and are interpreted as possibility distributions. For example, fuzzy values of local slope and altitude can be *low_slope, medium_height,* with membership functions μ_{low_slope} : [0°,90°] → [0,1] and $\mu_{medium\_height:}$ [0,9000] → [0,1] defined on the numeric values of slope and altitude respectively. An imprecise value *A* can be represented by an interval $[a_m, a_M]$ of reference numeric values, e.g. slope is [15°, 18°], with membership value $\mu_A(x)=1$ for $a_m \leq x \leq a_M$ and $\mu_A(x)=0$ otherwise. Imprecise values can be used to represent indeterminacy due to low resolution of the representation, or even to represent missing information.

In many real cases, the available data are precise but there could be uncertainty on their validity for several reasons: either because the source of the data cannot be completely trusted, or because one knows that the means of acquisition are not enough sophisticated and generate systematic errors; not least, because data are a result of a subjective analysis, such as surveyed data. Uncertainty on the validity of data can be represented by associating a trust score $t \in [0,1]$ with data, ex. slope is 30% with trust score *t*.

A compatibility relationship between fuzzy numeric values can be defined based on the fuzzy Jaccard similarity measure between fuzzy sets *A* and *B* defined on a continuous domain *D* as follows:

$$compatibility(A,B) = \frac{\int_{i \in A \cap B} \min(\mu_A(i), \mu_B(i))}{\int_{i \in A \cup B} \max(\mu_A(i), \mu_B(i))} \tag{1}$$

This definition computes a compatibility degree in [0,1] and is applicable also when either *A* or *B* or both are precise. In the case of discrete basic domain the integration is replaced by a sum.

When *A* and *B* are imprecise values $A=[a_m,a_M]$, $B=[b_m,b_M]$ their compatibility degree can be easily computed as follows:

$$compatibility(A,B) = \begin{cases} 0 & if\ a_M < b_m\ or\ b_M < a_m \\ \dfrac{min(a_M, b_M) - max(a_m, b_m)}{max(a_M, b_M) - min(a_m, b_m)} & otherwise \end{cases} \tag{2}$$

A distance measure between fuzzy or imprecise numeric values can also be defined as proposed in [12]. For imprecise values $A=[a_m,a_M]$, $B=[b_m,b_M]$ their distance is defined as:

$$\text{distance}(A,B) = \int_{-1/2}^{1/2} \left[\frac{(a_m + a_M)}{2} + x(a_M - a_m)\right] - \left[\frac{(b_m + b_M)}{2} + x(b_M - b_m)\right] \tag{3}$$

The distance between fuzzy values is derived by generalizing definition (3) (see [12]).

Imperfection of Ordinal Values

Ordinal values can be used in an imprecise way when one is unable to specify a single value of the ordinal domain (i.e. a label on an ordinal scale) but he/she can identify a set of values, e.g. a point in a map may be labeled as both *high* or *full* risk on the ordinal domain {*none, low, medium, high, full*}. Uncertainty on the validity on these values can be specified by associating with the data a trust score $t \in [0,1]$.

A compatibility degree can be computed for imprecise ordinal values $A=\{a_{min}, a_{Max}\}$and $B=\{b_{min}, b_{Max}\}$ defined on the ordinal scales $S_A=\{s_1,...,s_M\}$ and $Z_B=\{z_1,...,z_N\}$, respectively by applying definition (2) to their normalized set of indexes $p_A=[p_{a_{min}}, p_{a_{Max}}]$ and $p_B=[p_{b_{min}}, p_{b_{Max}}]$ computed as follows:

$$p_{s_k} = normalize\_index(s_k) = \frac{index(s_k) - index(s_{min})}{index(s_{max}) - index(s_{min})} \in [0,1] \quad (4)$$

with s_k ordinal value on a scale $S=\{s_{min},...,s_{Max}\}$ and

$$index(s_k)=argmax_{i=min,...k}(s_i)$$

The inverse function $normalize\_index^{-1}$ that maps a value $p \in [0,1]$ into and ordinal value s_i on a scale S with i=1,..., $|S|$, cardinality of S, is defined as:

$$s_i = normalize\_index^{-1}(p) = s_i \; with \left| p - \frac{index(s_i)}{2(|S|-1)} \right| = min_{j=1,..|s|} \left| p - \frac{index(s_j)}{2(|S|-1)} \right| \quad (5)$$

Imperfection of Nominal Values

Ambiguous categorizations derived by the inability to associate a single nominal value with a spatial position arise the need to define imprecise and uncertain nominal values. For example, this occurs in many applications of remote sensing when one has to classify a region into a vegetation type [24]. In these situations it can be useful to associate several vegetation types with the same spatial position thus defining a mixture type element such as v={0.8/*pine-forest* ; 0.6/*broadleaves-forest*}. These mixture values can be represented by fuzzy sets on the discrete basic domain of the original nominal types, ordered on a scale reflecting a numeric property, interpreted as possibility distribution. Thus, the compatibility degree between mixture type values can be defined by computing formula (1). A nominal value can be associated with a trust score representing this way its reliability or credit.

Imperfection of the Spatial Reference

Imperfection may also affect the spatial reference of a represented property, ex. vegetation type. For example, imperfection is introduced when rescaling an image to match a coarser resolution than the original one. To represent the spatial uncertainty of a property one can define the spatial reference implicitly associated with the pixel coordinates (i,j), through a fuzzy relation $R_{i,j}$:X×Y→[0,1] on the bi-dimensional spatial domain X×Y: ex. $R_{i,j}$ can be defined by the Cartesian product of two trapezoidal membership functions $(i_a,i_b,i_c,i_d)\times(j_a,j_b,j_c,j_d)$ with $i_a,i_b,i_c,i_d \in X$ and $j_a,j_b,j_c,j_d \in Y$. This way the grid becomes a fuzzy grid. The fuzzy relation value $R_{i,j}(x,y)$ in a pixel (i,j), ex. $R_{i,j}(x,y)=0.8$, expresses the possibility that the position (x,y) on the spatial domain has the value $v_{i,j}$, ex. $v_{i,j}$=*pine-forest*, while, at the same time, pixel (h,k) specifies the possibility degree $R_{h,k}(x,y)$, ex. $R_{h,k}(x,y)=0.6$, for the same point (x,y) to assume the value $v_{h,k}$, ex. $v_{h,k}$ = *broadleaves-forest.*

An alternative way for representing the uncertainty of the spatial reference is to use fuzzy pixels' values. For example, we can choose a precise spatial reference associated with the pixel coordinates (i,j) and a fuzzy value, ex. *v*={0.8/*pine-forest*; 0.6/ *broadleaves-forest*}, identifying a mixture type pixel. This way we express the fact that the precise cell on the spatial domain identified by the pixel coordinates (i,j) has the possibility to be either a *pine-forest* (with possibility degree 0.8) or a *broadleaves-forest* (with possibility degree 0.6). In the following we will adopt this second option: we assume a precise spatial reference associated with a pixel and incorporate uncertainty in representing the pixel value.

3 Consensual Fusion of Imperfect Values

The Fusion Framework

We have several decision maps represented by grids of pixels, possibly with imprecise or fuzzy values, generated by *n* competitive models, software tools, or human experts (the sources), each one characterized by a distinct trust score (representing its reliability, presumed credit), and we want to fuse their possibly contradictory values so as to achieve a more robust consensual decision map.

Let's consider n grids with the same spatial reference and resolution, in the following we indicate with $v_1,\ldots v_n$ the *n* values in the cell with indexes i,j of the *n* grids.

We assume that $v_1\ldots,v_n$ have the same basic domain, that can be either numeric discrete, numeric continuous, or ordinal. Further, each value v_i can be an imprecise value, i.e., an interval on the basic domain, or a fuzzy value, i.e., a convex possibility distribution.

We want to model the following fusion criteria:

- the greater the trust score of the source, the more the respective data must determine the consensual result;
- the greater the spatial agreement of a source within a specified neighborhood of each pixel with the other sources, the more the source contributes to determine the consensual result;
- the consensual result must be affected at most by a maximum uncertainty level specified by the decision maker;
- the fusion strategy should not be rigid and fixed once for all, but flexibly tunable depending on the needs of the application so as to model decision attitudes with distinct trade-offs between risk-taken and risk-adverse;

The mean like nature of fusion strategies has been outlined by many authors [10][13][14][17][20][31] and is recognized as particularly useful in the context of spatial decision making [2][8][9][15][19][22][23][25][26].

We represent the fusion strategy by modeling a decision attitude as a quantified-guided function by a monotone non decreasing linguistic quantifier Q defined by a fuzzy set μ_Q:[0,1]→[0,1] specified by a triple (a,b,c) with $a,b \in [0,1]$ and $c>0$ with the meaning depicted in Figure 1 [33] [31]

Q=all means that the pixel values in the consensual map must reflect the common decision of all the sources. In the case in which the values in the input maps are proportional to an alarm or anomaly condition, by specifying *all* one wants to model a risk-taken map: all the experts/models must agree on the need to issue the alarm on a given position of the map or to point at the anomaly in order to set an alarm in the consensual map for that position. *Q=at least 1* means that the pixel values in the consensual map must reflect the highest value. In the case in which the value is proportional to an alarm or anomaly, by selecting *at least 1* one models a risk-adverse map: one chooses the most alarming model. This can be useful in making precautionary decisions. *Q=most* means that the consensual map must reflect the shared decision of a fuzzy majority; this models a trade-off decision attitude between the two extreme cases.

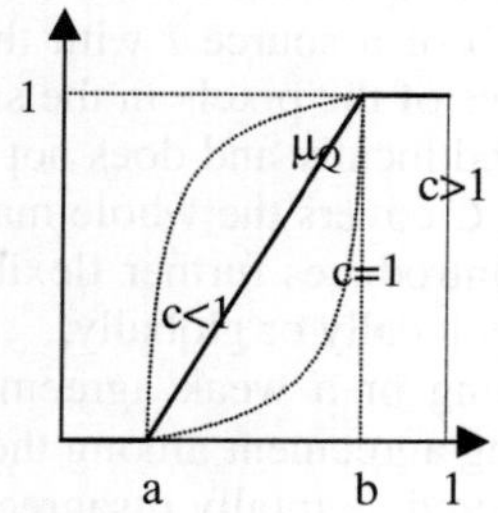

$$\mu_Q(x) = \begin{cases} 0 & x \le a \\ \left(\frac{x-a}{b-a}\right)^c & a < x < b \\ 1 & x \ge b \end{cases}$$

Fig. 1. Membership function of a relative monotone non decreasing linguistic quantifier specifying a fusion strategy.

Definition of the Fusion Attitude and of the Importance of the Sources

The fusion function associated with Q is defined by the generalized $\underline{\text{OWA}_Q}$ operator, defined below (in formula (7)) as an extension of the standard OWA operator [29] to aggregate possibly imprecise values, i.e., intervals on a real domain D.

The values to aggregate are weighted by importance degrees $i_1, \ldots, i_n \in [0,1]$, n number of sources, computed by taking into account both the trust scores of the sources, that we represent by values $t_k \in [0,1]$ with $k=1,\ldots,n$, and the agreements of the sources themselves defined by means of either a compatibility measure or a distance measure as follows :

$$i_i = \alpha * t_i + \beta * (Agreement(i,C)) \quad \text{with} \quad \alpha + \beta = 1 \quad \textit{and}$$

$$Agreement(i,C) = \frac{1}{|C|} \sum_{\forall (x,y) \in C} \frac{\sum_{k=1,k \neq i}^{n} f(v_i(x,y), v_k(x,y))}{\max_{i=1,\ldots,n} \sum_{k=1,k \neq i}^{n} f(v_i(x,y), v_k(x,y))} \quad (6)$$

$$\text{with} \qquad f = compatibility \quad \text{or}\, f = \max\_dist - distance$$

in which (x,y) are the coordinates of the pixels within the same subparts C of the maps, $|C|$ is the number of pixels in C, and f can be chosen as either a compatibility measure between values v_i,and v_j (*compatibility*(v_i,v_j) is defined as in (1) for fuzzy numeric values and as in (2) for imprecise values) or the complement of a distance measure as defined in (3) with *max_dist* being the maximum value of the distance on the considered domain of pixel values. The values in a map are important if they belong to a trusted map or, if they are in agreement with the correspondent values in the other maps. The parameters α and β control the relative influence of (1) the trust of a map, and of (2) the agreement degree, in the fusion operator. For example, by choosing β=1 and α=0, the influence of the values in the fusion strategy will be totally dependent on their agreement values.

Notice that the agreement degree *Agreement(i,C)* of a source *i* with the other *n*-1 sources is computed with respect to all the values of the pixels in the submaps *C*. If *C* is a single pixel (x,y) the agreement is defined locally and does not depend on the agreement of the other pixels in the maps. If *C* covers the whole maps then the agreement between the sources is global. This introduces further flexibility in the model since it allows considering data variability locally or globally.

The function *f* determines the choice for a strong or a weak agreement. By choosing a *compatibility* function we require a strong agreement among the values since two values having no overlapping are considered as totally disagreeing. By choosing a distance we are more tolerant of the differences among the values.

Definition of the Generalized OWA Operator for Imprecise Values

The weighting vector W_Q of the $\underline{\text{OWA}_Q}$ operator is derived by applying the following formula starting from the definition of μ_Q [32]:

$$w_i = \mu_Q\left(\frac{1}{e}\sum_{j=1}^{i} e_j\right) - \mu_Q\left(\frac{1}{e}\sum_{j=0}^{i-1} e_j\right) \qquad \textit{with} \quad e = \sum_{i=1}^{n} e_i = \sum_{j=1}^{n} i_j \qquad (7)$$

where e_j is the importance degree i_j computed by applying formula (6) associated with the j-th largest value to aggregate. This way w_k, i.e., the increment in satisfaction in having *k* non-null values with respect to *k-1,* increases with e_k. The values with no importance play no role. The data from the sources with greatest agreement with other sources and highest trust score determines more heavily the increment in satisfaction and then is more heavily taken into account by the fusion function.

The generalized $\underline{\text{OWA}}_Q$ operator of dimension n and weighting vector W_Q, with $\sum_{i=1,...n} w_i=1$, and w_i computed by formulae (6) and (7), aggregates n imprecise values $[v_{1,m},v_{1,M}]\ldots,[v_{n,m},v_{n,M}]$, $v_{1,m},\ldots v_{1,M}\ldots,v_{n,m},\ldots v_{n,M}\in D$ (D is a continuous domain), and computes an imprecise value $[c_{1,m},c_{1,M}]$ of D. This operator is defined as follows:

$$\underline{\text{OWA}}_Q : R(D)^n \rightarrow R(D)$$

where $R(D)$ is the set of all intervals on D and:

$$[c_{1,m},c_{1,M}] = \underline{\text{OWA}}_Q([v_{1,m},v_{1,M}],\ldots,[v_{n,m},v_{n,M}])$$

$$\underline{\text{OWA}}_Q([v_{1,m},v_{1,M}],\ldots,[v_{n,m},v_{n,M}])=\sum_{i=1,\ldots n} w_i*[g_{i,m}, g_{i,M}] \quad (8)$$

in which $[g_{i,m}, g_{i,M}]$ is the i-th largest interval of the $[v_{1,m},v_{1,M}],\ldots, [v_{n,m},v_{n,M}]$ such that:

Order: $[a_1,a_2] > [b_1, b_2]$ if $(a_1+a_2)/2 > (b_1+b_2)/2$
Addition: $[a_1, a_2] + [b_1, b_2] = [a_1+b_1, a_2+b2]$
Product: $[a_1, a_2] * [b_1, b_2] = [a_1* b_1, a_2* b_2]$

Application of the Generalized OWA to Different Kind of Data

When all values to be aggregated are precise the $\underline{\text{OWA}}_Q$ reduces to the usual OWA_Q definition [29]. If the values to fuse are defined on a discrete domain D we have to apply a further rounding function to the result of $\underline{\text{OWA}}_Q([v_{1,m},v_{1,M}],\ldots,[v_{n,m},v_{n,M}])$ so as to yield an interval $[c_{1,m},c_{1,M}]$ defined on the same discrete domain D. In the case in which the data to fuse are ordinal values, several proposals have been defined in the literature for the definition of the fusion operation [11][14]. In our approach, we apply the fusion operation defined in (8) to the intervals $[p_{1,m},p_{1,M}],\ldots,[p_{n,m},p_{n,M}]$ defined by the normalized indexes $p_{1,m},p_{1,M},\ldots,p_{n,m},p_{n,M}\in[0,1]$ derived by applying formula (4) to the imprecise ordinal values. The result is then an interval of [0,1] that must be turned back into an imprecise ordinal value defined on the same scale of the original data by applying the function defined in (5). Finally, in the case in which the values to fuse are fuzzy values, represented by convex possibility distributions μ_v, we apply the $\underline{\text{OWA}}_Q$ operator to their u–cuts(μ_v)=$\{x \mid \mu_v(x) > u\}$, where u is the maximum uncertainty level (specified by the decision maker) that can be tolerated in the consensual result. In fact, if we apply an u–cuts to a possibility distribution representing some real variable we can say that the values in the u–cut are affected by at most an uncertainty degree equal to u, i.e., we cannot be completely sure that the real value of the variable is in the set u–cut, unless u=0. Thus, if we apply the fusion to the u–cuts affected by an uncertainty u we obtain a fused imprecise value affected at most by the same uncertainty degree.

7 Experiments: Seismic Hazard Analysis

As a first example of application of the consensual fusion strategy we discuss the generation of a consensual seismic ground motion map based on the fusion of six maps produced independently by applying distinct input models.

The six computed ground motion maps are referred to the same area (Calabria region, Southern Italy) and each one is associated with a trust score (in this case a ground motion value *g*, which is a positive real number). In the classical approach the fused map is generated as the weighted average of the maps in which the weight is the trust score [21]; the trust is computed by using a logic tree. As a matter of fact, in standard Probabilistic Seismic Hazard Analysis (PSHA) logic trees explicitly characterize the epistemic uncertainties (trusts) residing in the adopted models. Logic trees offer a clear representation of the probabilities of alternative choices on fundamental parameters and models concurring tothroughby a weighted mean of all the alternative computed hazard values.

In figure 2 we have depicted the six maps of ground motions independently computed by the six input models. The grey level represents the difference between the ground motion values of the first map (high on the left) produced by the first model with respect to the others. It can be noticed that the models mostly disagree in their estimation of the ground motion values in the central area of the maps.

The proposed consensual fusion function defined in (8) is applied to generate the consensual ground motion map relative to a specified fuzzy majority of the trusted models.

We applied our approach by modeling two distinct fusion strategies: a risk-taken fusion defined by the quantifier *most* (a=0.6, b=0.9, c=1 – See Figure 1) and a risk-adverse fusion defined by the quantifier *some* (a=0.0, b=0.3, c=1).

In our approach, we take into account the imprecision of the models in generating their ground motion maps by representing the pixel values through fuzzy numbers (g_m, g, g_M) in which g is the ground motion value computed by the model in the current pixel, and g_m, $<g<$ g_M are defined to capture the approximations applied by the models. The imprecise values of ground motion to fuse by applying definition (8) are derived as 0-cuts of the fuzzy numbers of ground motion, i.e., $0\text{-}cut((g_m, g, g_M)) = [g_m, g_M]$. This way we require maximum certainty on the fused result.

In computing the importance of a value from a source by applying formula (6) we considered a local definition of the agreement among the models: this way the agreement of a model with the others is computed independently for each pixel of the fused map.

Figure 3.a and 3.b depict the maps obtained by the difference of the two consensual fusion strategies specified by *most* and *some* with respect to the classic weighted mean.

It can be observed that the most precautionary strategy corresponding with *some* in figure 3.b produces as expected higher ground motion values than the weighted mean while the opposite occurs for the most risk-taken strategy specified by *most* depicted in figure 3.a.

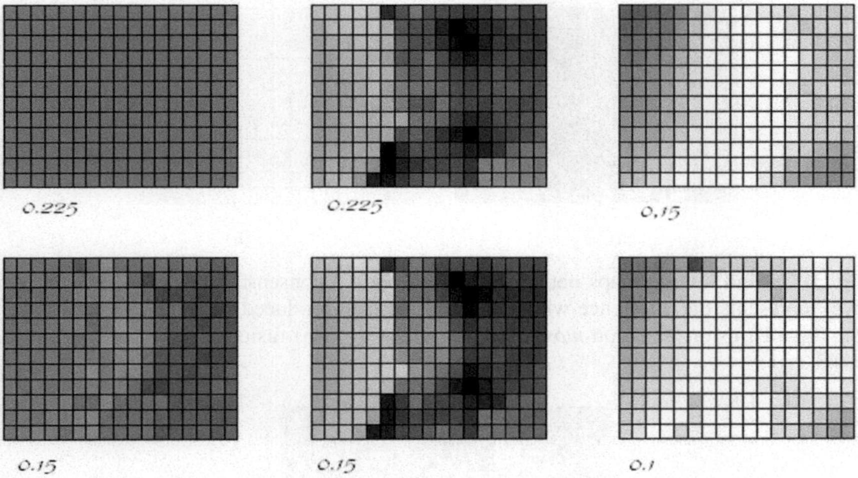

Fig. 2. Ground motion maps computed by the six models: the trust score is indicated below each map. The grey level represents the difference of the ground motion value with respect to the first map on the left.

To show the influence of the consensual dynamics on the results we show in Figure 4 the map obtained by the difference of the classic weighted mean map and the consensual map corresponding with the quantifier *average* (a=0,b=1,c=1) that models an average (arithmetic mean) of all the models. It can be observed that the effect of the consensual dynamics is more evident in the central region of the map where there is the lower agreement among the original maps in figure 2. Specifically, in this area only two out of the six models determine high ground motion values, while the other four models agree for lower values. The consensual ground motion values in this area are then in accordance with the majority of the models, i.e., the ground motion values are lower with respect to those produced by the classic approach.

A second experiment based on the application of the proposed consensual fusion was for the estimation of consensual iso-probable response spectra in seismic hazard analysis [34]. In particular, we have applied the consensual fusion aggregation described in this chapter and compared its results to the results produced by the classic approach.

We performed a PSHA for a village placed along the Po river (Northern Italy). The scheme of the analysis followed the structure of the logic-tree defined by INGV in 2004 for the computation of the seismic hazard of the Italian national territory [35].

In our analysis we used two alternative models for the assessment of catalogue completeness, two models for the computation of the occurrences and five alternative ground motion prediction equations. Figure 6 shows the 20 isoprobable spectra (5% damping) characterized by an exceedance probability of 10% and 2% in 50 years respectively.

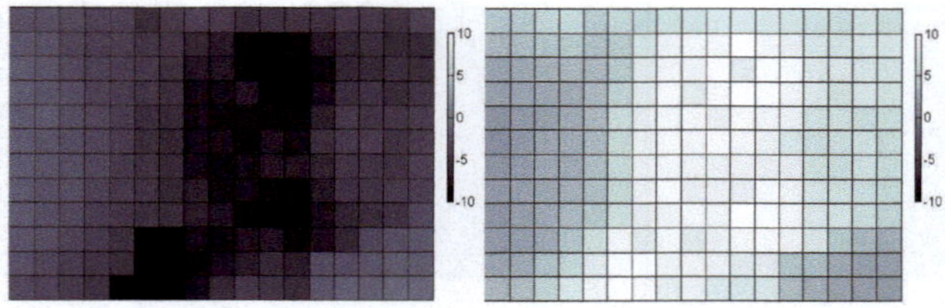

(a) (b)

Fig. 3. Ground motion maps obtained by the proposed consensual fusion model: the grey level represents the difference with respect to the map produced with the classic weighted average. (a) fusion based on *most* (a=0.6, b=0.9, c=1); (b) fusion based on *some* (a=0.0, b=0.3, c=1).

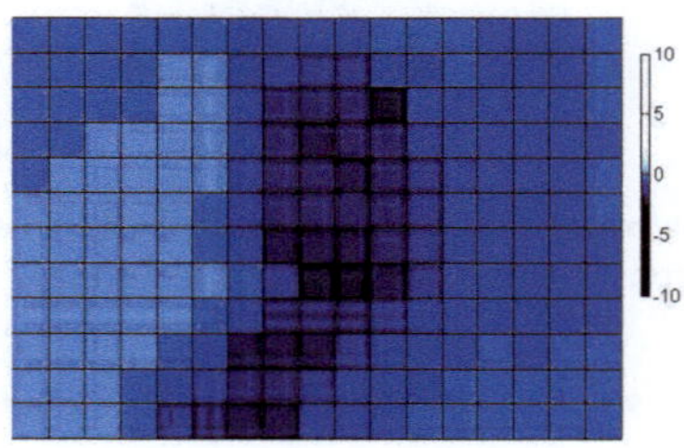

Fig. 4. differences between the ground motion maps obtained by the consensual fusion based on *average* (a=0.0, b=1.0, c=1.0) and the classic weighted mean.

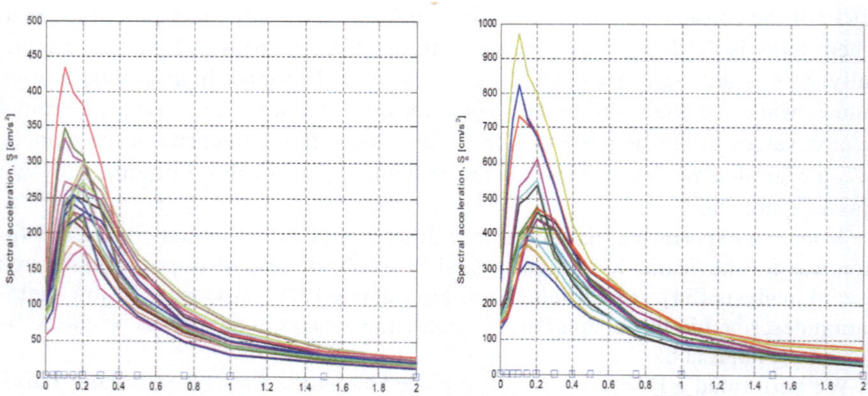

(a) **(b)**

Fig. 5. The **(a)** panel shows the isoprobable response spectra with 10% probability of exceedance in 50 years—5% damping, computed for the test-site, while the **(b)** panel reports the isoprobable spectra with 2% probability of exceedance in 50 years—5% damping, for the same site.

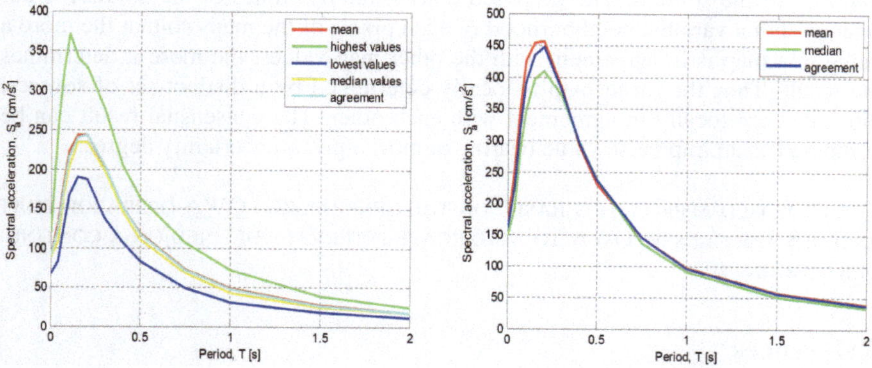

(a) (b)

Fig. 6. The left **(a)** panel shows single iso-probable spectra (10% probability of exceedance in 50 years—5% damping) computed for the test-site using different fusion strategies: classic approach based on mean and median estimates, mean of the highest (at least a few, close to OR fusion) and lowest values (almost all, close to AND fusion), and consensual mean fusion. The right **(b)** panel show single isoprobable spectra (5% probability of exceedance in 50 years—5% damping) computed for the test-site using mean, median and consensual mean fusions.

Successively we applied the aggregation of these isoprobable spectra by first using the OWA aggregation operator without consensus with several attitudinal vectors, i.e., linguistic quantifiers, that corresponds to the classical statistical operators mean and median, and to the max and min aggregations.

Besides this experiment we applied the consensual fusion with the same attitudinal vectors, mean and median, by considering the consensus among the isoprobable spectra. In this respect we considered a local agreement. The trust weights have been set as the product of the weights on each path of the tree from the root to the leafs. The fused spectra are shown in figure 7. The few examples we have shown illustrate some interesting aspects especially in the case of the 2475yr. Isoprobable response spectra aggregation (see Figure 7(b)). In this situation the consensual mean fusion gives different results with respect to the ones obtained with a classical mean estimate.

5 Conclusions

In this chapter we analyzed the aspects involved in the fusion of imperfect multisource spatial data. We first considered the problem of representing the imperfection of spatial data in the fuzzy set framework. The proposed fusion model is applicable to multisource spatial data affected by imperfection. The model takes into account several aspects of uncertainty in the fusion: the trust of the sources, the imprecision and uncertainty of the spatial data values, and the vague nature of the fusion strategy that is represented by a linguistic quantifier identifying a fuzzy majority of the sources. The proposed consensual dynamics of the sources is defined within a variable neighborhood of each pixels of the maps so that the more a value in a map is in agreement with the other map values, the more it determines the result. Thus the fused map is locally determined by a distinct set of sources, those that are locally in agreement with each other. The consensual result can be computed as an imprecise value bearing at most a given uncertainty degree.

THE PROPOSED APPROACH IS BASED ON THE GENERALIZED OWA OPERATOR THAT EXTENDS THE CLASSIC OWA TO AGGREGATE INTERVALS DEFINED ON A CONTINUOUS DOMAIN.

References

[1] Bloch, I., Maître, H.: Information combination operators for data fusion: A comparative review with classification. IEEE Transactions on Systems, Man, and Cybernetics, Part A: Systems and Humans 26(1), 52–67 (1996)

[2] Bone, C., Dragicevic, S., Roberts, A.: Integrating high resolution remote sensing, GIS and fuzzy set theory for identifying susceptibility areas of forest insect infestations. International Journal of Remote Sensing 26(21), 4809–4828 (2005)

[3] Bordogna, G., Chiesa, S., Geneletti, D.: Linguistic modelling of imperfect spatial information as a basis for simplifying spatial analysis. Information Sciences 176, 366–389 (2006)

[4] Bordogna, G., Pagani, M., Pasi, G.: A Flexible Decision support approach to model ill-defined knowledge in GIS. Presented at the NATO Workshop on Environmental Impact Assement, Kiev (June 2006)

[5] Bosc, P., Prade, H.: An Introduction to the Fuzzy Set and Possibility Theory-based Treatment of Flexible Queries and Uncertain or Imprecise Databases. In: Motro, A., Smets, P. (eds.) Uncertainty Management in Information Systems, pp. 285–324. Kluwer, Dordrecht (1997)

[6] Brivio, P.A., Boschetti, M., Carrara, P., Stroppiana, D., Bordogna, G.: Fuzzy integration of satellite data for detecting environmental anomalies across Africa. In: Hill, J., Roeder, A. (eds.) Advances in Remote Sensing and Geoinformation Processing for Land Degradation Assessment. Taylor & Francis, London (2006)

[7] Burrough, P.A., McDonnel, R.A.: Principles of Geographical Information Systems. Oxford University Press, Oxford (1998)

[8] Chanussot, J., Mauris, G., Lambert, P.: Fuzzy fusion techniques for linear features detection in multitemporal sar images. IEEE Trans. on Geoscience and Remote Sensing 37(3), 1292–1305 (1999)

[9] DeCETI Project, Multi-sources information fusion for satellite image classification, electronic Report of the DeCETI Project, Leonardo da Vinci Programme, Strand II, Measure II.1.1.C, Contract No: GR/1996/II/0953/PI/II.1.1.c/FPC (2000), http://www.survey.ntua.gr/main/labs/rsens/DeCETI/IRIT/MSI-FUSION/ (accessed 20/12/2006)

[10] Dubois, D., Prade, H.: Possibility theory and data fusion in poorly informed environments. Control Engineering Practice 2(5), 811–823 (1994)

[11] Gogo, L., Torra, V.: On Agrregation Operators for Ordinal Qualitative Information. IEEE Trans. on Fuzzy Systems 8(2), 143–153 (2000)

[12] Gu, Q.P., Cao, B.Y.: Approach to Linear Programming with Fuzzy Coefficients Based on Fuzzy Numbers Distance. In: FUZZ IEEE 2005, pp. 447–450 (2005)

[13] Herrera, F., Herrera-Viedma, E., Martnez, L.: A fusion approach for managing multi-granularity linguistic terms sets in decision making. Fuzzy Sets Syst. 114(1), 43–58 (2000)

[14] Herrera, F., Herrera-Viedma, E.: Linguistic decision analysis: Steps for solving decision problems under linguistic information. Fuzzy Sets Syst. 115(1), 67–82 (2000)

[15] Jiang, H., Eastman, J.R.: Application of fuzzy measures in multi-criteria evaluation in GIS. International Journal of Geographical Information Science 14(2), 173–184 (2000)

[16] Kacprzyk, J.: Group decision making with a fuzzy linguistic majority. Fuzzy Sets and Systems 18, 105–118 (1986)

[17] Kam, M.: Performance and geometric interpretation for decision fusion with memory. IEEE Trans. on Systems, Man, and Cybernetics, Part A 29(1), 52–62 (1999)

[18] Keenan, P.B.: Spatial Decision Support Systems: An coming of age. Control and Cybernetics 35, 9–27 (2006)

[19] Malczewski, J.: GIS-based multicriteria decision analysis: a survey of the literature. International Journal of Geographical Information Science 20(7), 703–726 (2006)

[20] Morris, A., Jankowski, P.: Fuzzy techniques for multiple criteria decision making in GIS. In: Joint 9th IFSA World Congress and 20th NAFIPS International Conference, Vancouver, CA, July 25-28, pp. 2446–2451 (2001); (CD ROM proceedings)

[21] Rabinowitz, N., Steinberg, D.M., Leonard, G.: Logic Trees, sensitivity analysis and data reduction in probabilistic seismic hazard assessment. Earthquake Spectra 14(1), 189–201 (1998)

[22] Robinson, P.B.: A perspective on the fundamentals of fuzzy sets and their use in Geographic Information Systems. Transactions in GIS 7(1), 3–30 (2003)

[23] Silvert, W.: Ecological impact classification with fuzzy sets. Ecological Modelling 96, 1–10 (1997)

[24] Solaiman, B.: Multisensor data fusion using fuzzy concepts: application to land-cover classification using ERS-1/JERS-1 SAR composites. IEEE Trans. on Geoscience and Remote Sensing 37(3), 1316–1326 (1999)

[25] Stroppiana, D., Boschetti, M., Brivio, P.A., Carrara, P., Bordogna, G.: Continental monitoring of vegetation status with a fuzzy anomaly indicator: an example for Africa. Submitted to Remote Sensing of Environment (2006)

[26] Tran, L.T., Knight, C.G., O'Neill, R.V., Smith, E.R., Riitters, K.H., Wickham, J.: Environmental assessment, fuzzy decision analysis of integrated environmental vulnerability assessment of the Mid-Atlantic region. Environmental Monitoring 29(6), 845–859 (2002)

[27] Valet, L., Mauris, G., Bolon, P.: A statistical overview of recent literature in information fusion. IEEE AESS Systems Magazine 1, 7–14 (2001)

[28] Wald, L.: Some terms of reference in data fusion. IEEE Trans. on Geoscience and Remote Sensing 37(3), 1190–1193 (1999)

[29] Yager, R.R.: On ordered weighted averaging aggregation operators in multi-criteria decision making. IEEE Trans. on Systems, Man and Cybernetics 18, 183–190 (1988)

[30] Yager, R.R.: Quantifier guided aggregation using OWA operators. International Journal of Intelligent Systems 11, 49–73 (1996)

[31] Yager, R.R.: A framework for multi-source data fusion. Information Sciences 163, 175–200 (2004)

[32] Yager, R.R.: Interpreting Linguistically Quantified Propositions. International Journal of Intelligent Systems 9, 541–569 (1994)

[33] Zadeh, L.A.: A computational approach to fuzzy quantifiers in natural languages. Computers and Mathematics with Applications 9, 149–184 (1983)

[34] Pagani, M., Pagani, M., Bordogna, G., Marcellini, A.: About the use of the OWA operator in case of a PSHA based on a logic tree, xxx

[35] di Lavoro, G.: Redazione della Mappa di pericolosità sismica prevista dall'ordinanza PCM 3274 del 20 marzo 2003. Rapporto conclusivo per il Dipartimento di Protezione Civile, INGV, Milano-Roma, 65pages +5 appendixes (April 2004), `http://zonesismiche.mi.ingv.it/`

Part II

Database Querying, Spatial and Temporal Databases

Querying Fuzzy Spatiotemporal Databases: Implementation Issues

Aziz Sözer*, Adnan Yazıcı, Halit Oğuztüzün, and Frederick E. Petry

Abstract. Modeling and querying spatiotemporal data, in particular fuzzy and complex spatial objects representing geographic entities and relations are challenging topics that have many applications in geographic information systems. In a recent article the authors have presented an approach to these problems. The present chapter focuses on the issues that arise from implementing this approach. As a case study the implementation of a meteorological database application that combines an object-oriented database with a knowledgebase is discussed.

1 Introduction

Spatiotemporal applications involve space and time related data. Hence, database systems are required to deal with both spatial and temporal phenomena. The application domains are numerous (e.g. traffic control, cadastral, meteorological and environmental information systems). Such applications typically include attribute variations, primarily due to moving objects. For example a moving car in city traffic changes position over time. Another source of variations is the changing attributes of objects such that the borders of a salty lake move back and forth because of seasonal evaporation and rainfall. Combining these two phenomena, some applications include moving objects with changing attributes. Meteorological event

Aziz Sözer, Adnan Yazıcı, and Halit Oğuztüzün
Department of Computer Engineering, Middle East Technical University, Ankara, Turkey

Frederick E. Petry
Naval Research Laboratory Code 7440.05, Bldg 1005 Stennis Space Center, MS 39529
e-mail: e070364@ceng.metu.edu.tr
* Correspondence author.

J. Kacprzyk et al. (Eds.): Uncer. Appro. for Spatial Data Model. and Process., SCI 271, pp. 97–116.
springerlink.com

monitoring yields many such examples: On a weather map, a stormy area moves while possibly changing its shape. Thus, spatiotemporal data handling requires advanced data structures and modeling techniques [16, 18].

Another property of most spatiotemporal applications is fuzziness, because geometric and topological properties usually involve various forms of uncertainty. For example, in describing a windy region on a weather chart, the region's boundary is fuzzy. The rivers change line because of floods and drought [8]. In the case of estimating a moving weather object, the need to determine its position at a certain time, or its time of arrival at a certain location, give rise to fuzzy estimations. It is not always easy to obtain precise data, and we may only be able to give a range of values in which the exact numbers would lie. For instance, we may need the number of *cloudy* or *partly cloudy* days for some region in a period. Instead of giving numeric degrees of cloudiness (e.g. 4/8) linguistic terms can be used [2]. These facts lead researches to leveraging fuzzy set theory for modeling spatial objects and their properties [20, 27]. Schneider [20] represents fuzzy spatial objects and relationships as well as complex crisp objects and relationships by using fuzzy techniques. Clementini [4] introduces a geometric model for uncertain lines which is a basis for the study of topological relations.

Temporality has also been studied by some researchers [19, 28]. In its simplest form, time is considered as an attribute of spatial objects. A simple time stamping approach is adequate to obtain the states of objects at certain times. However, to identify individual changes in objects, event-based approaches are developed [19]. In [28] temporal uncertainty and fuzzy timing are introduced to a model that combines temporality and fuzziness. This model features the concepts of fuzzy time stamping, enabling time, occurrence time and delays.

There are efforts to combine spatial and temporal properties into one modeling framework [9, 10]. In [9], an object-oriented modeling approach which is very useful for modeling and manipulating spatiotemporal data and having unique features (e.g. encapsulation, polymorphism, dynamic binding, aggregation, etc.) is used. Yazici et. al. [26] used unified modeling language (UML) [1], providing extensions to handle spatial and temporal objects. In their work, some new special entity sets, relationships, and constructs were introduced for modeling spatial objects. Fuzzy object-oriented modeling techniques are used to model and analyze the imperfect information requirements of various complex applications [13, 15, 25]. Marin et. al. [15] present a set of operators useful to compare objects in a fuzzy setting. Among them are a generalized resemblance degree between two fuzzy sets of imprecise objects and a generalized resemblance degree to comp are complex fuzzy objects. Yazici et. al. [25] have studied a similarity based fuzzy object-oriented data model in which impreciseness at the data level contributes to uncertainty in the class-object level and that in class-subclass hierarchy. In this paper we introduce some extensions to that model for spatiotemporal objects.

In knowledge intensive applications, support for deduction is an essential requirement. In a spatiotemporal application, relations between objects can be very complex. Consider the following, for example: There is a ship crossing the sea and in some parts the sea line may be restricted for travel due to wave and wind conditions. How can we record this information and make the deduction that the sea line is restricted? A knowledgebase that is capable of making deductions and providing knowledge would be very helpful for retrieving the status of the sea line. Hence, the interaction and/or integration of database and knowledgebase technologies are important requirements for the development of knowledge intensive applications. This is reflected in the continuing research into the development of deductive object-oriented models since the late 1980s [5].

In this study, we revisit our new approach to model and query real world spatiotemporal objects, in particular meteorological phenomena [22]. We combine our fuzzy object-oriented database model with a knowledgebase to cope with the deduction requirements of the application. The specific contribution of [22] comes with a generic modeling of spatiotemporal database applications and a fuzzy spatiotemporal querying mechanism. The generic data model developed in [22] has been implemented as a proof-of-concept application. In addition, crisp/fuzzy spatial/nonspatial querying, which may require inferencing, is handled by utilizing the Intelligent Fuzzy Object-Oriented Database (IFOOD) architecture [12]. We give some information on concepts related to fuzzy spatiotemporal database modeling, including spatial and temporal fuzziness as well as relationships between fuzzy and complex objects in Background section. In Section 3, we describe how to develop a generic model for spatiotemporal applications. We use a meteorological application to illustrate our approach. Section 4 gives details about the architectural design of the system. In Section 5 we present our implementation details together with example queries from the application domain, and discuss crucial details of their processing. Finally, in the last section, we present our conclusions and point to possible future studies.

2 Background

In this section spatial and temporal concepts are discussed. Specifically, spatiotemporal objects, attributes, operations and relations are defined.

2.1 Spatial Data Types

In geographic information systems, natural and man-made objects (e.g. mountains, aridity areas, cadastral divisions, roads and meteorological phenomena like foggy regions, wavy sea regions, etc.) are modeled and queried. The objects are defined with spatial (e.g. shape, location, boundary length, diameter etc.) and/or descriptive (e.g. name, population etc.) attributes [14]. In Figure 1, the *wave heights* over the Mediterranean Sea are illustrated on a weather chart.

In the weather chart, the wave heights have varying characteristics, for example, they are most dense in south-west of Italy and clear on the Eastern Mediterranean. The borders of the density regions are indeterminate since their characteristics

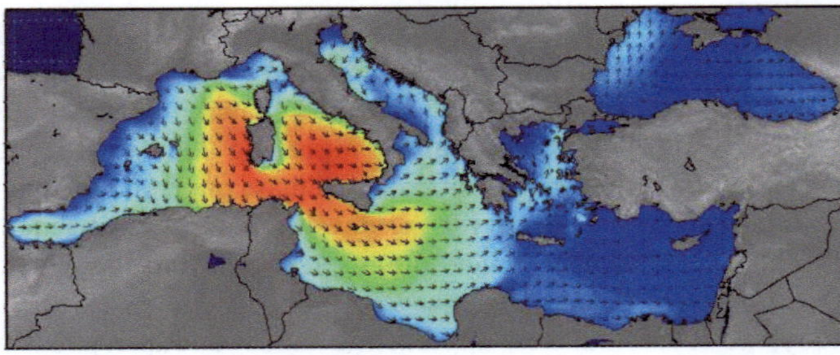

Fig. 1. A weather chart showing wave heights on the Mediterranean Sea.

change somewhat gradually. In a geographic space like this chart, the objects with imprecise or vague spatial attributes (e.g. wave height) could be referred to as *fuzzy spatial objects* and the ones with precise or exact attributes (e.g. country borders) could be referred to as *crisp spatial objects* [20].

Hereafter, we shall first give the definitions for fuzzy spatial objects, which are fuzzy points, lines and regions. Then we will define fuzzy/complex crisp relations. A *fuzzy point* is a point for which an exact position is not known but possible positions are known to be within a certain area. In Figure 2-(a) the expected position of such a point is shown by a black dot and the possible positions are shown by grey dots. For instance, a ship waiting in the queue for crossing *Istanbul Bosphorus* is found at a certain point but may change its position from time to time (e.g. move to the grey parts). A *fuzzy line* is a line, the exact shape, position or length of which is not known, but it is known in which area the line must be. In Figure 2-(b) the center line shows the normal shape of a river. The actual river line can change position and shape due to floods or droughts (e.g. the grey area). A *fuzzy region* has three parts: (1) the core (the dark part) (2) the indeterminate boundary (the grey part) and (3) the exterior (the outer parts of indeterminate boundary) [27]. In Figure 2-(c) a typical fuzzy region is depicted and might be used to express the gradual change over a spatial domain for a given attribute (e.g. wave height). Finally, a *complex region* is as set of regions, possibly with holes and multiple components (see, Figure 2-(d)). Foggy regions with clear patches or cloudy regions with rainy parts can be represented as complex regions.

(a) (b) (c) (d)

Fig. 2. Spatial Data Types.

2.2 Spatial Relations between Objects

Spatial relationships are divided into three: topological (e.g., *overlap*), directional (e.g., *North of*) and metric (e.g., *5 km away from*) relationships [3]. Topological relations describe spatial intersection or relationships of objects in space. A model for analyzing binary topological relations, known as the *9-intersection model*, has been proposed in the literature [7]. The 9-intersection model is based on the intersection between the parts (*interior, boundary and exterior*) of the regions involved. The intersections of the parts are analyzed with 3x3 matrices (total 2^9=512 matrices). The model distinguishes eight meaningful (*disjoint, meet, overlap, equal, contains, inside, covers* and *covered by*) relations for crisp regions. Later, this model was generalized for fuzzy regions [17, 23, 27] and complex regions [6, 20].

2.3 Fuzzy Topological Relations

Suppose that *A* is a set of attributes under consideration, and that a region is a fuzzy subset defined in two-dimensional space R^2 over *A*. The membership function of the region can be defined as $\mu : X \times Y \times A \rightarrow [0,1]$, where *X* and *Y* are the sets of coordinates defining the region. Each point (x, y) within the region is assigned a membership value for an attribute *a* in *A*. A fuzzy region is illustrated in Figure 3 with the *core*, *the indeterminate boundary*, *exterior* and *α-cut levels.* The concept of $\alpha - cut$ level region is used to approximate the indeterminate boundaries of a fuzzy region and defined as follows:

$$R_\alpha = \{(x, y, a) | \mu_R(x, y, a) \geq \alpha\}(0 < \alpha < 1) \quad (1)$$

Next, fuzzy topological relations between fuzzy regions are defined [23]. Fuzzy topological relations are inevitably fuzzy because of the indeterminate boundaries

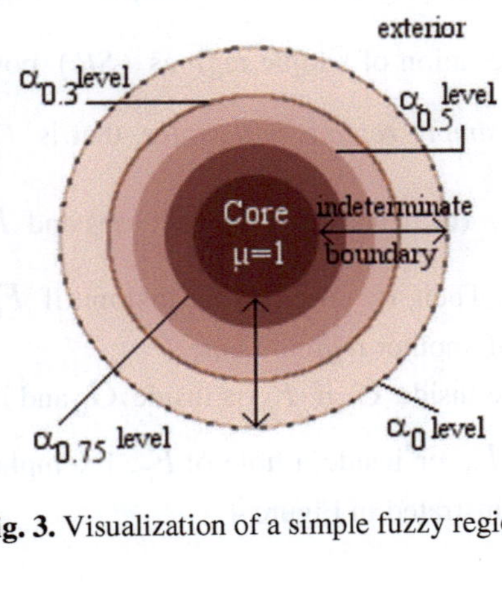

Fig. 3. Visualization of a simple fuzzy region.

of the regions involved. The degree of the fuzzy relation is measured by aggregating the $\alpha - cut$ level of fuzzy regions. The *basic probability assignment* $m(R_{\alpha i})$, which can be interpreted as the probability that $R_{\alpha i}$ is the true representative of R , is defined as in [20, 27]:

$$m(R_{\alpha i}) = \alpha_i - \alpha_{i+1}, 1 \leq i \leq n, n \in N, 1 = \alpha_1 > \alpha_2 > \ldots > \alpha_n > \alpha_{n+1} = 0 \quad (2)$$

It is clear that $\sum_{i=1}^{n} m(R_{\alpha i}) = 1$.

Let $\tau(R, S)$ be the value representing the topological relation between two fuzzy regions *R* and *S*, and $\tau(R_{\alpha i}, S_{\alpha j})$ be the value representing the topological relation between two $\alpha - cut$ level regions $R_{\alpha i}$ and $S_{\alpha j}$. Then the general relation between two fuzzy regions can be determined by

$$\tau(R, S) = \sum_{i=1}^{n} \sum_{j=1}^{m} m(R_{\alpha i}) m(S_{\alpha j}) \tau(R_{\alpha i}, S_{\alpha j}) \quad (3)$$

For example, the overlap relation between two fuzzy regions can be approximated by using the formula above as follows:

$$\tau(R, S) = \sum_{i=1}^{n} \sum_{j=1}^{m} m(R_{\alpha i}) m(S_{\alpha j}) \tau_{overlap} (R_{\alpha i}, S_{\alpha j}) \quad (4)$$

The remaining topological relations can be analyzed in a similar manner.

2.4 Topological Relations between Complex Regions

A complex region is the union of simple regions (SR) possibly including holes. Let F and G be two *simple regions with holes*, that is $F_{SR} = F_0 - \bigcup_{i=1}^{n} F_i$ (5) and $G_{SR} = G_0 - \bigcup_{j=1}^{m} G_j$ (6), where F_0, G_0 are bases and F_i, G_j are the holes of F and G respectively. Then, two regions are disjoint if F_0 and G_0 are disjoint or one region is inside of another region's hole.

F is considered to be inside G if F_0 is inside G_0 and if each hole G_j of G is either disjoint from F_0 or inside a hole of F_i. Exemplary regions with holes and their relations are illustrated in Figure 4.

(a) (b)

Fig. 4. Examples of the relations (a) disjoint and (b) inside.

Other topological predicates for simple regions are defined in the same vein. Next, based on these definitions, topological predicates for complex regions are defined as follows:

Let $F_{CR} = \bigcup_{i=1}^{n} F_i$ (7) and $G_{CR} = \bigcup_{j=1}^{m} G_j$ (8) be two *complex regions* (CR), where F_i and G_j are simple regions with holes. Then topological relations for complex regions are defined as follows:

$$disjoint_{CR}(F,G) \Leftrightarrow \forall 1 \le i \le n\ \forall 1 \le j \le m : disjoint_{SR}(F_i, G_j)$$

$$inside_{CR}(F,G) \Leftrightarrow \forall 1 \le i \le n\ \exists 1 \le j \le m : inside_{SR}(F_i, G_j)$$

$$meet_{CR}(F,G) \Leftrightarrow \neg disjoint_{CR}(F,G) \wedge (\forall 1 \le i \le n\ \forall 1 \le j \le m : (disjoint|meet)_{SR}(F_i, G_j))$$

$$contains_{CR}(F,G) \Leftrightarrow inside_{CR}(G,F)$$

$$equal_{CR}(F,G) \Leftrightarrow n = m \wedge \forall 1 \le i \le n : equal_{SR}(F_i, G_i)$$

$$coveredBy_{CR}(F,G) \Leftrightarrow \neg((inside|equal)_{CR}(F,G)) \wedge (\forall 1 \le i \le n\ \exists 1 \le j \le m : (inside|coveredBy|equal)_{SR}(F_i, G_j))$$

$$covers_{CR}(F,G) \Leftrightarrow coveredBy_{CR}(G,F)$$

$$overlap_{CR}(F,G) \Leftrightarrow \neg(disjoint \mid meet \mid inside \mid contains \mid equal \mid coveredBy \mid covers)_{CR}(F,G)$$

2.5 Temporal Requirements

Temporal aspects have been the focus of attention in the literature, and applications frequently require that these be captured in the database. Information about objects' attributes and relationships among objects are valid *when* the object exits temporally. For example, windy regions exist over the sea within a time interval and the ships which have to cross these regions are planned to start and finish their journeys at certain times. The windy regions and the ship routes will be expected to relate to each other in certain ways in this interval.

To handle temporal aspects, time is generally stored in databases in two forms:

(a) *Valid time* is the time when the information about an object or relationship holds in the modeled reality. For example the valid times of a ferry route in the Marmara Sea is 08:30, 12:00 and 17:00 daily.

(b) *Transaction time* of a database entry is the time when the entry is part of the current state of the database. The time when the ferry lines' times are stored in the database is the transaction time of the entry.

3 A Generic Model for Spatiotemporal Querying

In this section, the components of a generic spatiotemporal model, which is a fuzzy object oriented database (FOOD) [25] and a fuzzy knowledgebase (FKB) [12] are presented.

3.1 The Fuzzy Object-Oriented Database (FOOD) Model

The Fuzzy Object Oriented model supports multivalued attributes for which fuzzy domains are defined. The *domain* of an attribute is the set of all possible values that the attribute can take. For example, the fuzzy domain for a "temperature" attribute of a meteorological observation can be defined as:

$$Domain_{temperature} = \{\text{hot, warm, moderate, cool, cold}\}$$

That is, the temperature attribute can have some combination of these values from the domain such as {hot, warm}, {warm}, {cool, cold, moderate}. In FOOD, attributes are defined within a *range* which is a set of allowed values that the attribute can take. In general, $range \subseteq domain$. The range of an attribute a_i of a class C is represented by the notation $rng_c(a_i)$, where $a_i \in \{a_1, a_2, \ldots, a_n\}$, *the attributes of class C.* For example, the range of the temperature attribute of a class for a *fog* object can be defined as a subset of the temperature domain:

$$rng_{fog}(temperature) = \{moderate, cool, cold\}$$

The similarity matrix in Table 1 shows the similarity of each element with other elements in the domain. The matrix indicates that cool and cold temperatures are similar with a degree of 0.8. In a case where the temperature value is estimated only and given a threshold value of 0.8 or lower, multiple values {cool, cold} can be associated, which gives us a fuzzy representation for temperature value.

Another type of fuzziness in FOOD occurs between classes and objects. That is, while some objects are full members of a fuzzy class, some other objects may belong to the class partially. The objects may still be considered as instances of this class but with a degree of membership in [0, 1]. The degree of membership of an object to its class is computed by using the similarities between the attribute values, the class range values and the relevance of fuzzy attributes [25].

Table 1. The similarity matrix for temperature attributes

Temperature	hot	warm	moderate	cool	cold
hot	1.0	0.6	0.4	0	0
warm	0.6	1.0	0.8	0.2	0
moderate	0.4	0.8	1.0	0.6	0.4
cool	0	0.2	0.6	1.0	0.8
cold	0	0	0.4	0.8	1.0

3.2 The Object Model

In this paper, we consider a meteorological application where the objects and relations are modeled with the extended UML [26]. The object model presented in Figure 5 consists of two parts: the first part includes generic spatial and temporal classes that can be used by any specific application domain and the second part consists of meteorological application classes.

The spatial part consists of *ST_Geometry*, *Point*, *Line* and *Region* classes. These classes are denoted as part classes and the relationship in between is illustrated by the aggregation constructor (denoted by a diamond symbol). A special form of aggregation (i.e. the "whole/part" relation) exists between *ST_Geometry*, the whole, and the part classes, which is indicated by a double diamond symbol. In the whole/part relation the whole is aggregated by different kinds of parts. The μ attribute in spatial classes stores a membership value to describe a proximity to a certain fixed space. So, the objects may belong to a class fully (i.e. with a degree of 1) or partially (i.e. with a membership degree larger than zero and less than or equal to one). As an extension to UML, a fuzzy class constructor, indicated by a double-square placed on the upper-left hand side of the spatial class, explicitly represents the fuzzy instances. The other fuzzy constructor, indicated by the tag *U* to the left-hand side of the name of the class, is used to indicate the existence of class attributes having uncertain values, such as the size of fuzzy geometries.

The temporal dimension is represented by *DateTime* class. *UType* as an attribute type, indicates an uncertain type such that the attribute value may belong to a domain of valid values, and may be *null*, *incomplete*, *non applicable* or expressed with a level of precision. The spatial and temporal dependency of a class is shown by *S and T* on the upper right-hand side of the entities.

The *Fuzzy* class provides range definitions, relevance values and class-object membership values for *ST_Object*. Under the generic model, meteorological application classes (*MetObject, GeoLine*, etc.) inherit *ST_Object*. The *MetObject* class represents a meteorological object. The model is completed with other class definitions such as a *GeoLine* refers to a geographic line object which may be a real one like a river, a railway, or a virtual line like the route of a ferry or a plane.

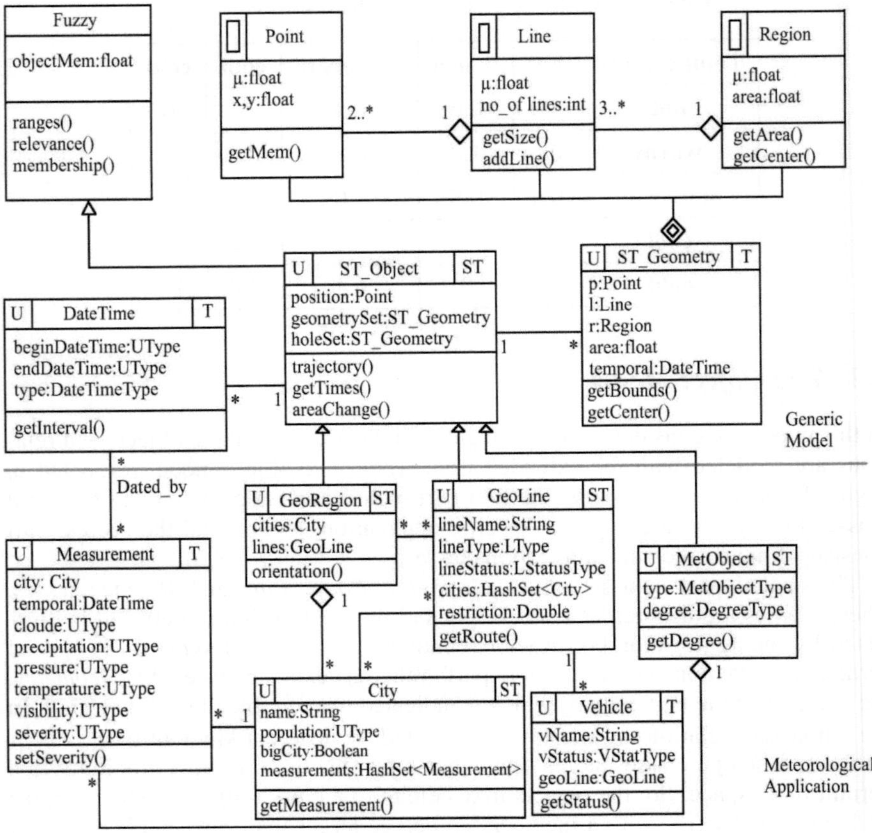

Fig. 5. A fuzzy spatiotemporal model

3.3 Exemplary Class Definitions

In this section exemplary class definitions are presented to complete the object model. *ST_Object* class is defined as follows:

```
public abstract class ST_Object extends Fuzzy{
    ST_Geometry gset[]; //simple geometries
    ST_Geometry hset[]; //holes
    int ngset, nhset; //number of geometries and holes
    ST_Point position; //Center of the object
    Temporal[] times; //temporal entries
    float[] sizes; //Size of the object
    float[] orientation; //Directions of the object
    ST_Point[] trajectory(); //Position list
}
```

The trajectory of the object is an ordered sequence of points for ordered times. While object is following a trajectory, we also hold in parallel size changes (*sizes*) and orientation values (*orientation*).

3.4 Coupling the Fuzzy Database with a Fuzzy Knowledgebase

In order to achieve an intelligent application, a knowledgebase (*KB*) is integrated to the object-oriented database. We utilize the *Intelligent Fuzzy Object Oriented Database (IFOOD)* [12], which provides flexible and powerful querying mechanisms for complex data and knowledge with uncertainty in both database and knowledgebase.

The KB used in the IFOOD architecture includes rules and intelligent objects having fuzzy attributes. It features a fuzzy inference method used for deduction of fuzzy conclusions, gets the rules and objects as input, tries to satisfy rules by comparing with facts, and produces a conclusion from the satisfied rules.

4 The Architecture of the Spatiotemporal Database Application

The architecture of the proposed environment for spatiotemporal data modeling is illustrated in Figure 6. The user interface (*UI*) component gets user inquires and sends these query constraints to the bridge interface (*BI*). After the query is completed the query results are displayed to the user textually and/or graphically. The *BI* component plays a coordinating role in query processing. The communication and interaction between the database system, the knowledge base system and the fuzzy spatial processor is performed by the BI. It gets user queries, analyzes them, sends requests to the database and/or to the knowledge base, retrieves the results, and sends them up to the user interface.

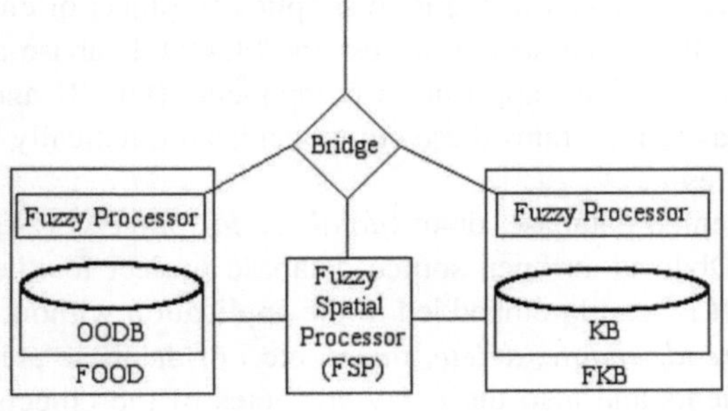

Fig. 6. The architecture of the spatiotemporal database application.

The *FOOD* system acts as a database server for objects. The definitions of uncertain types, similarity relations, and membership functions are stored in the object-oriented database. Fuzzy processors are used to handle uncertainty at both the database component and the knowledgebase component so that users are able to inquire objects having uncertain properties and probably firing some rules in the knowledgebase within the same query.

The fuzzy knowledgebase (*FKB*) system processes rules taking fuzzy objects as input. We provide the required facilities in the FKB system to access the definitions in the FOOD system. For example, if the FKB system needs the similarity of two fuzzy terms of some domain, it gets this value via the fuzzy processor from the FOOD system.

The fuzzy spatial processor (*FSP*) module processes topological predicates between complex crisp and fuzzy spatial objects. BI forwards the user request to FSP if the query includes a topological predicate. FSP requests the spatial objects from FOOD and finds the predicates and the degree of membership of the relation.

5 Implementation

The application makes use of the following technologies and tools:

- *Application development environment:* NetBeans IDE 6.1
- *Java:*1.6.0_07; Java HotSpot™ Client VM 10.0-b23
- *System:* Windows XP version 5.1 running on x86; Cp 1254.
- *Object Oriented Database System:* db4o-6.4.44.10817-java5.jar
- *Knowledge Base:* jess.jar

We used the NetBeans Integrated Development Environment (*IDE*) for application development. The NetBeans IDE is open-source and supports development of Java applications. The graphical user interface of the IDE saves the time by managing ready made components (forms, buttons, panels, etc.), settings and data.

The architecture presented in Figure 6 supplies an object oriented database and a knowledgebase. It is necessary to access the FOOD database and fuzzy knowledgebase components from application components (UI, BI and FSP) and vice versa. The IDE easily integrates these components automatically by adding corresponding ".jar" files.

The object oriented database, db4o (*database for objects*) [24] is embedded in the application. Db4o is an open source database project for the object-oriented database model. It is readily embedded in the application without any installation. The operations (*read, update, delete, insert*, etc.) of database are used by the application components and also the fuzzy attributes of the objects are handled by means of our application.

As a rule engine and scripting environment, *Jess* [11] is embedded in our application. In Jess, we can reason about objects using the knowledge supplied in the form of declarative rules. Here is an exemplary rule:

```
(deftemplate GeoLine (declare (from-class GeoLine)))
(defglobal ?*fp* = (new Fsp))
(defrule geolinestatus
?pl <- (GeoLine
    (lineType ?lT&:(and (neq ?lT nil) (eq ?lT "SeaLine")))
    (threshold ?th)
    (OBJECT ?obj))
    =>
(bind ?result (call ?*fp* FuzzyRelation ?obj "wave" "wavy"))
(bind ?result2 (call ?*fp* FuzzyRelation ?obj "wind" "windy"))
(bind ?minresult (min ?result ?result2))
(if (> ?minresult ?th) then
  (call ?obj setlineStatus "restricted"))
(if (< ?minresult ?th) then
  (call ?obj setlineStatus "clear"))
(call ?obj setOverlap ?minresult))
```

If the supplied object of the class *GeoLine* has an overlap degree with meteorological objects *wave* and *wind* larger than or equal to a threshold value then line status is set to *restricted*. Jess uses the *Rete* algorithm to process rules [21]. Its scripting language allows us to access the complete Java's API so that we can create Java objects, call Java methods and implement Java interfaces.

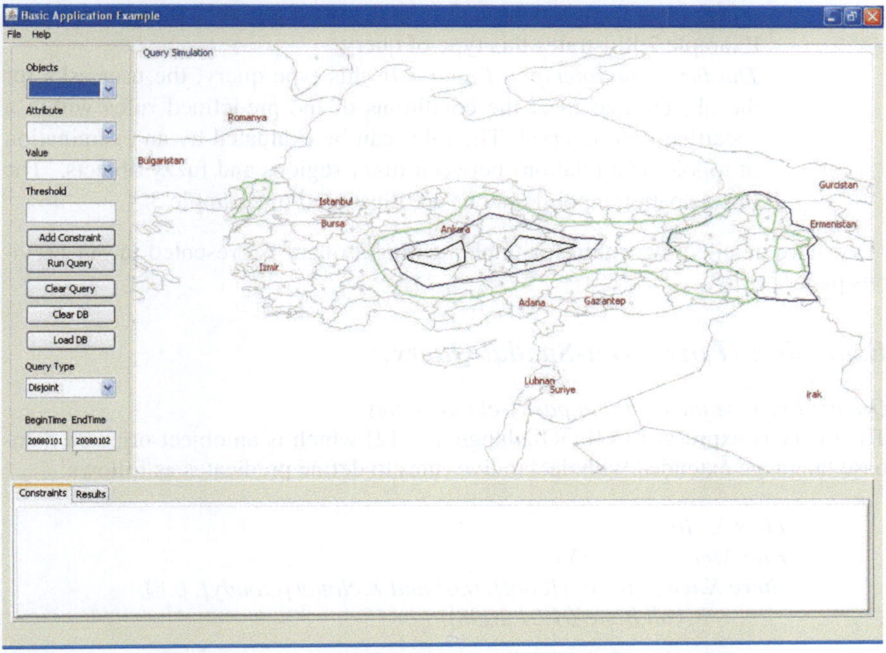

Fig. 7. A sample screenshot of Meteorological Application.

In Figure 7, a sample screenshot of the application is depicted. As the sample data, the geographic objects (e.g. cities, seas, etc.) inside and around Turkey are used. The meteorological object data (e.g. temperature, precipitation, etc) for some days are obtained from Turkish Meteorological Service to illustrate how the system works and for the proof of concept. For performance evaluation and efficient querying much bigger data sets should be stored in the database with proper indexing mechanisms, which is our future study.

The following procedures are applied to resolve the query according to its type:

- *The basic query (crisp and non-spatial)*: This type of query asks for crisp data that does not have a spatial dimension. For instance the basic attributes of *City* objects such as name, population, etc. or the measurements data. This type of query is parsed and sent to OODB directly and the results are displayed by UI.
- *The fuzzy non-spatial query*: This type of query asks for data that is fuzzy but non-spatial and the BI, FKB, and OODB components are employed. The objects retrieved by the BI are sent to the FKB component to check whether they meet the fuzzy conditions. How these objects are checked is illustrated in Example 1. Objects satisfying the conditions are sent back to the BI.
- *The complex spatial query*: Complex spatial objects and their relationships are queried in this type of query. The BI, OODB and the FSP components are employed to fetch query results. The user asks for the objects that have topological relations with the objects under inquiry. Example 2 illustrates this type of query.
- *The fuzzy spatiotemporal query*: In this type query, the user asks for the objects that meet the conditions of the predefined rules within a specified time interval. The rules can be evaluated by an examination of topological relations between fuzzy regions and fuzzy objects. The fuzzy spatiotemporal queries are illustrated in Example 3.

A more detailed algorithm for implementing a query is presented in our previous paper [22].

Example 1 (Fuzzy Non-Spatial Query):

Query: *Retrieve the cool and partly cloudy cities.*
The query is expressed in IFOOD language [12] which is an object-oriented database language extended with declarative rules to define predicates as follows:

select *X.city*
from *Measurement(X)*
where *X.temperature([cool], 0.6) and X.cloud([cloudy], 0.8),*
X.validtime(01.01.2008);

Table 2. Sample *Measurement* objects in database

ID	City	Cloud	Temperature	Severity
C1	Istanbul	cloudy	cool	normal
C2	Edirne	partly cloudy	moderate	normal
C3	Izmit	cloudy, closed	cold	severe

Table 3. Similarity matrix of cloud attribute

Cloud	clear	partly cloudy	cloudy	closed
clear	1.0	0.6	0.1	0
partly cloudy	0.6	1.0	0.5	0.1
cloudy	0.1	0.5	1.0	0.4
closed	0	0.1	0.4	1.0

The objects used in the example are listed in Table 2, and the similarity relation of *cloud* is included in and Table 3. The similarity relation of temperature is already presented in Section 3.1.

The query is evaluated as follows:

i The first predicate to evaluate in this query is *X.temperature([cold],0.6)*.

- C1.temperature is cool, and $\mu_{Similarity}(cool, cool)=1.0$. Therefore C1 satisfies the temperature predicate.
- C2.temperature is moderate, and $\mu_{Similarity}$ $(cool, moderate)=0.6$. Therefore C2 satisfies.
- C3.temperature is cold, and $\mu_{Similarity}$ $(cool, cold)=0.8$. Therefore C3 satisfies.

ii Then, the predicate *X.cloud ([cloudy], 0.8)* is evaluated.

- C1.cloud is cloudy, and $\mu_{S\ Similarity}$ $(cloudy, cloudy) =1.0$. Therefore C1 satisfies the cloud predicate.
- C2.cloud is partly cloudy, and $\mu_{Similarity}(cloudy, partly\ cloudy)=0.5$. Therefore C2 does not satisfy.
- C3.cloud is cloudy or closed with $max\{\mu_{Similarity}(cloudy, cloudy), \mu_{Similarity}(cloudy, closed)\}= max\{1.0, 0.4\}=1.0$. Therefore C3 satisfies.

iii As a result, the objects C1 and C3 satisfy the fuzzy query conditions.

Example 2 (Complex Spatial Query):

In Figure 8-a the maximum temperature regions and in Figure 8-b the meteorological events are illustrated as mapped by the Turkish Meteorological Service for 01, January 2008.

The temperature regions are shown in different colors (e.g. cold parts by dark blue, cool parts by green, moderate parts by orange and warm parts by red). Temperature regions are visualized as complex spatial objects since they have multiple

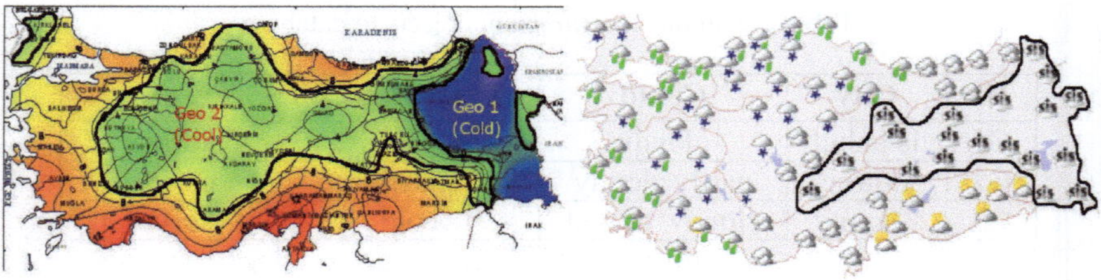

(a) (b)

Fig. 8. Maximum temperature regions (a) and meteorological events (b).

components. The expected meteorological events are depicted with symbols and colors, e.g. rain (green drops), snow (blue stars), grey clouds, black foggy areas and yellow patchy areas. We assume the meteorological objects representing temperature regions, fog, precipitation and cloud are inserted in the database as shown in Table 4. The temperature regions, which are classified by their degrees (e.g. cool, cold, etc.), have different geometries with multiple components (e.g. Geo_1, Geo_2, etc.). According to the figure, Cold (dark blue) region has one simple region (Geo_1) and the cool regions (green) have four simple regions.

Table 4. Objects in the FOOD

Object	Type	Degree	Geometry Set	Valid Time
MetObject	Temperature	cold	$\{Geo_1\}$	01.01.2008
MetObject	Temperature	cool	$\{Geo_2, Geo_3, Geo_4, Geo_5\}$	01.01.2008
MetObject	Fog	foggy	$\{Geo_6\}$	01.01.2008
MetObject	Precipitation	rainy	$\{Geo_7, Geo_8\}$	01.01.2008
MetObject	Precipitation	snow	$\{Geo_9\}$	01.01.2008
MetObject	Cloud	cloudy	$\{Geo_{10}\}$	01.01.2008
MetObject	Cloud	partlycloudy	$\{Geo_{11}\}$	01.01.2008

Query: *Retrieve the cold and foggy regions on 01, January 2008.*
This query is expressed in IFOOD as follows:

select *X.geometry, Y.geometry, spatialRelation(X,Y)*
from *MetObject(X), MetObject(Y)*
where *X.type([temperature])* ***and*** *Y.type([fog])* ***and*** *X.degree([cold],0.8)* ***and*** *Y.degree([foggy],0.8)* ***and*** *X.validtime(01.01.08)* ***and*** *Y.validtime(01.01.08);*

In this query, the temperature objects having the attribute value *cold,* and the fog objects having the *foggy* degree are fetched from FOOD to BI. The user supplies a threshold value 0.8 for temperature degree, so cool regions are also fetched since $\mu_s([cold], [cold]) = 1.0$ and $\mu_s([cold], [cool]) = 0.8$. The simple topological relation algorithm is applied for components of complex regions. After finding simple topological predicates, the complex topological relation algorithm is applied to determine the final topological predicate [22].

Example 3 (Fuzzy Spatiotemporal Query):

In this example, fuzzy spatial relations are queried. In Figure 9, wave height and wind speed for *Marmara Sea* are illustrated at 31.12.2007 15:00 Greenwich Mean Time (GMT) (between 40.0-41.4 North latitudes and 26-30 East longitudes). According to the figure the central parts have the highest waves and strongest winds while the coastal areas have lower waves and calm winds. The three lines, namely, *Line*$_1$, *Line*$_2$ *and Line*$_3$ represent certain ferry routes between the ports.

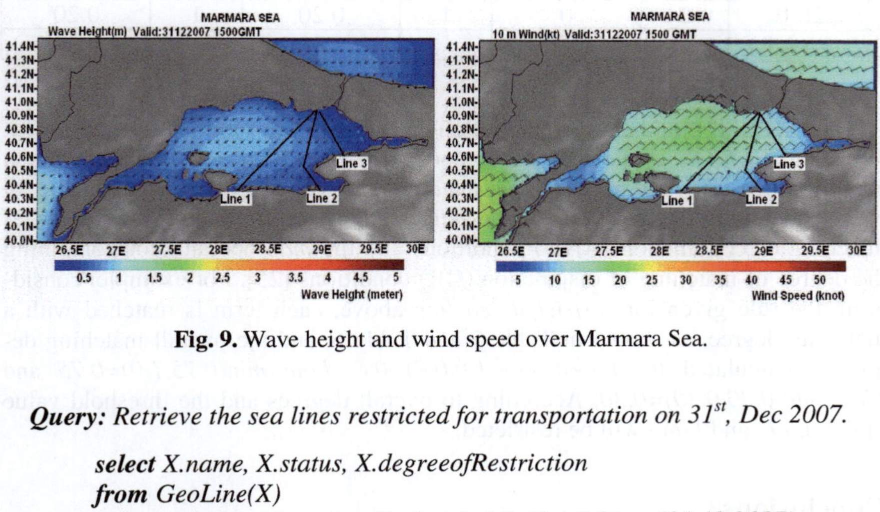

Fig. 9. Wave height and wind speed over Marmara Sea.

Query: *Retrieve the sea lines restricted for transportation on 31*[st]*, Dec 2007.*

> ***select*** *X.name, X.status, X.degreeofRestriction*
> ***from*** *GeoLine(X)*
> ***where*** *X.status([restricted], threshold), X.validtime (31.12.2007);*

In the query, the sea lines that are restricted more than a given threshold value are requested. The geographic line status is a rule defined in the FKB as defined in section 5. The required objects (sea wind and wave height geometries) are fetched from the OODB, and FSP calculates the fuzzy spatial relation (overlap in this case) between the *fuzzy regions* wind and wave, and *crisp* ferry lines using the *fuzzy topological relation algorithm* [22].

According to the meteorological forecast, the sea area is divided into *five* α *– cut levels* and the ferry lines overlap some of them (see Figure 9); the calculation details are presented in Table 5 and Table 6.

Table 5. Computing a fuzzy topological relation for a wavy region and ferry lines

α-cut level (wave)	$\tau_{overlap}$	m(region)x m($line_1$)	$\tau_{overlap}$	m(region)x m($line_1$)	$\tau_{overlap}$	m(region)x m($line_1$)
1.0-0.75	1	0.25	0	0	0	0
0.75-0.50	1	0.25	1	0.25	0	0
0.50-0.30	1	0.20	1	0.20	0	0
0.30-0	1	0.30	1	0.30	1	0.30
$\tau_{overlap}$(R, L)		1.0		0.75		0.30

Table 6. Computing a fuzzy topological relation for a windy region and ferry lines

α-cut level (wind)	$\tau_{overlap}$	m(region)x m($line_1$)	$\tau_{overlap}$	m(region)x m($line_1$)	$\tau_{overlap}$	m(region)x m($line_1$)
1.0-0.65	1	0.35	1	0.35	0	0
0.65-0.30	1	0.35	1	0.35	0	0
0.30-0.20	1	0.10	1	0.10	1	0.10
0.20-0	0	0	1	0.20	1	0.20
$\tau_{overlap}$(R, L)		0.80		1.00		0.30

The results of the fuzzy spatial relation calculations are supplied to FKB for inferencing. In FKB, a rule may be composed of more than one condition. Each condition in a rule may have its own matching degree. Therefore, we compute an overall matching degree. We use the *min* operator for combining the degree of matching of conjunction (AND) conditions and the *max* operator for combining the degree of matching of disjunction (OR) conditions [25]. For example, considering the rule given for *restricted sea line* above, each term is matched with a matching degree, as shown in Table 5 and Table 6, and the overall matching degree is calculated for $Line_1$: $min(1.0,0.8)=0.8$, $Line_2$:$min(0.75,1.0)=0.75$ and $Line_3$:$min(0.30,0.30)=0.30$. According to overall degrees and the threshold value of 0.7, $Line_1$ and $Line_2$ will be restricted.

Conclusions

In this study we have presented a generic spatiotemporal data model and a querying mechanism for spatiotemporal databases. We presented our method, designed to handle uncertainty in spatiotemporal database applications. We used an application, involving meteorological objects with some spatial and temporal attributes, as an example. The proposed mechanism has been implemented as a proof-of-concept prototype. We have discussed several implementation issues that arise.

In the scope of this work, meteorological phenomena and geographic data are modeled as spatiotemporal objects. These objects can move and evolve in time. In addition, the meteorological and geographic man made objects may have spatial relations. The crucial decision was to integrate the model with a fuzzy knowledgebase allowing a fuzzy deduction and querying capability to handle complex data and knowledge. As a result, we are able to handle spatiotemporal queries (position, spatial properties and spatial relationships).

Spatiotemporal data modeling and querying require further research. The model and the method presented in this paper should be applied to other fields, such as wireless sensor networks and multimedia, to gain more insight into fuzzy spatiotemporal modeling and querying. Another future research topic is the development of efficient indexing mechanisms for fuzzy spatiotemporal databases.

Acknowledgements

Adnan Yazıcı would like to thank TUBITAK (The Scientific and Technological Research Council of Turkey) for sponsoring this research under grant EEEAG-106E012. Fred Petry would also like to thank the Naval Research Laboratory's Base Program, Program Element No. 0602435N for sponsoring this research.

References

[1] Booch, G., Rumbaugh, J., Jacobson, I.: The Unified Modeling Language User Guide Reading. Addison-Wesley, Reading (1999)

[2] Bordogna, G., Chiesa, S., Geneletti, D.: Linguistic modelling of imperfect spatial information as a basis for simplifying spatial analysis. Information Sciences 176, 366–389 (2006)

[3] Claramunt, C., Theriault, M.: Fuzzy semantics for direction relations between composite regions. Information Sciences 160, 73–90 (2004)

[4] Clementini, E.: A model for uncertain lines. Journal of Visual Languages & Computing 16(4), 271–288 (2005)

[5] Colomb, R.M.: Deductive Databases and Their Applications. Taylor & Francis, Abington (1998)

[6] Du, S., Qin, Q., Wang, Q., Ma, H.: Evaluating structural and topological consistency of complex regions with broad boundaries in multi-resolution spatial databases. Information Sciences 178, 52–68 (2008)

[7] Egenhofer, M., Clementini, E., Felice, P.: Topological relations between regions with holes. International Journal of Geographical Information Systems 8(2), 129–144 (1994)

[8] Fisher, P., Arnot, C., Wadsworth, R., Wellens, J.J.: Detecting change in vague interpretations of landscapes. Ecological Informatics 1, 163–178 (2006)

[9] Frihida, A., Marceau, D.J., Theriault, M.: Spatiotemporal object-oriented data model for disaggregate travel behavior. Transactions in GIS 6(3), 277–294 (2002)

[10] Griffiths, T., Fernandes, A., Paton, N., Barr, R.: The TRIPOD spatio-temporal data model. Data and Knowledge Engineering 49(1), 23–65 (2003)

[11] Jess, the Rule Engine for the Java Platform, http://herzberg.ca.sandia.gov/jess/

[12] Koyuncu, M., Yazici, A.: IFOOD: An Intelligent Fuzzy Object-Oriented Database Architecture. IEEE Trans. on Knowledge and Data Engineering, 1137–1154 (2003)

[13] Lee, J., Xue, N.L., Hsu, K.H., Yang, S.J.: Modeling imprecise requirements with fuzzy objects. Information Sciences 118, 101–119 (1999)
[14] Lu, C.T., Kou, Y., Zhao, J., Chen, L.: Detecting and tracking regional outliers in meteorological data. Information Sciences 177, 1609–1632 (2007)
[15] Marin, N., Medina, J.M., Pons, O., Vila, M.A.: Object resemblance in a fuzzy object-oriented context. In: Proceedings of 2002 IEEE International Conference on Fuzzy Systems, Honolulu, EEUU (2002)
[16] Nievergelt, J., Widmayer, P.: Spatial data structures: Concepts and Design Choices. In: Handbook of Computational Geometry, pp. 725–764 (2000)
[17] Pauly, A., Schneider, M.: Topological predicates between vague spatial objects. In: Bauzer Medeiros, C., Egenhofer, M.J., Bertino, E. (eds.) SSTD 2005. LNCS, vol. 3633, pp. 418–432. Springer, Heidelberg (2005)
[18] Pelekis, N., Theodoulidis, B., Kopanakis, I., Theodoridis, Y.: Literature review of spatio-temporal database models. The Knowledge Engineering Review 19(3), 235–274 (2004)
[19] Peuquet, D.: Making space for time: Issues in Space-Time Data Representation. GeoInformatica 5(1), 11–32 (2001)
[20] Schneider, M.: A design of topological predicates for complex crisp and fuzzy regions. In: Kunii, H.S., Jajodia, S., Sølvberg, A. (eds.) ER 2001. LNCS, vol. 2224, p. 103. Springer, Heidelberg (2001)
[21] Sosnowski, Z.A.: Activation of Fuzzy Rules in RETE network. In: Proc. of the 4th International Conference on Flexible Query Answering Systems (FQAS 2000), Warsaw, Poland. Advances in Soft Computing, pp. 200–209. Physica-Verlag (2000)
[22] Sozer, A., Yazici, A., Oguztuzun, H., Tasci, O.: Modeling and querying fuzzy spatiotemporal databases. Information Sciences (2008), doi:10.1016/j.ins.2008.05.34
[23] Tang, X., Fang, Y., Kainz, W.: Fuzzy topological relations between fuzzy spatial objects. In: Wang, L., Jiao, L., Shi, G., Li, X., Liu, J. (eds.) FSKD 2006. LNCS (LNAI), vol. 4223, pp. 324–333. Springer, Heidelberg (2006)
[24] http://www.db4o.com Db4o: Database for Objects
[25] Yazici, A., George, R., Aksoy, D.: Design and Implementation Issues in the Fuzzy Object-Oriented Data (FOOD) Model. Information Sciences 108(4), 241–260 (1998)
[26] Yazici, A., Zhu, Q., Sun, N.: Semantic data modeling of spatiotemporal database applications. International Journal of Intelligent Systems 16, 881–904 (2001)
[27] Zhan, F.B., Lin, H.: Overlay of two simple polygons with indeterminate boundaries. Transactions in GIS 7(1), 67–81 (2003)
[28] Zhou, Y., Murata, T.: Petri net model with fuzzy timing and fuzzy-metric temporal logic. International Journal of Intelligent Systems 14(8), 719–745 (1999)

Bipolar Queries: A Way to Deal with Mandatory and Optional Conditions in Database Querying

Sławomir Zadrożny and Janusz Kacprzyk

Abstract. We discuss an approach to bipolar queries. We start with the original idea proposed in Lacroix and Lavency and review some selected relevant approaches recently proposed in the literature. In particular we point out two main lines of research, the one focusing on a formal representation within some well established theories and the analysis of a meaningful combinations of multiple conditions, and the one concerned mainly with the study of how to aggregate mandatory (negative, or required) and optional (positive, or desired) conditions. We follow the second line of reasoning and show some relations with other approaches, both concerning database querying, exemplified by Chomicki's queries with preferences, and Yager's works in multicriteria decision making. In the former case we offer a fuzzy counterpart of a new relational algebra operator *winnow* and show how a bipolar query can be represented via the select and *winnow* operators.

1 Introduction

One of main challenges in the present day IT is a growing discrepancy between the computer and the human being. Basically, the power of the computer – both in the sense of hardware and, maybe to a lesser extent, software – is constantly growing at a rapid speed, while the "power" of the human being has been probably the same for the last centuries as the human information processing capabilities are presumably not growing. This gap between the human being and the computer has many aspects, and for our purposes the most important may be an *articulation/communication gap*.

Sławomir Zadrożny
Systems Research Institute, Polish Academy of Sciences,
ul. Newelska 6, 01-447 Warszawa, Poland
e-mail: Slawomir.Zadrozny@ibspan.waw.pl

Janusz Kacprzyk
Systems Research Institute, Polish Academy of Sciences,
ul. Newelska 6, 01-447 Warszawa, Poland
e-mail: Janusz.Kacprzyk@ibspan.waw.pl

J. Kacprzyk et al. (Eds.): Uncer. Appro. for Spatial Data Model. and Process., SCI 271, pp. 117–132.
springerlink.com

Its essence is mainly implied by the fact that for the computer the only fully natural means of articulation and communication are strings of 0s and 1s while for the human being this role is played by natural language with its inherent imprecision.

These gaps between the human being and the computer have far reaching consequences for all aspects of IT because the human being is a key element of virtually all nontrivial present IT systems and applications. Ways to overcome these gaps have been objects of intensive research for many years, and though much progress has been made, there are still many unsolved problems.

One of main elements of virtually all IT projects and applications are database management systems. An ability to make an effective and efficient use of information stored in a database(s) is a prerequisite for success. Unfortunately, the traditional means of database querying do not take into account specifics of the human being as mentioned above, notably - which is relevant for our purposes – they do not make the use of natural language possible in a simple and straightforward way, as it would have been preferred by the human user.

For many years there have been numerous attempts to overcome the above difficulty, notably to somehow make querying languages (exemplified by the most popular SQL) more human friendly and human consistent. This has resulted in the appearance of a large, new field in database research, called *flexible queries*, and the famous series of conferences FQAS (Flexible Query Answering Systems) has been biannually held for the last years (cf. many papers in the list of references to this paper).

An interesting area within that field of flexible querying are fuzzy queries in which through the use of natural language terms (values of some attributes, linguistic quantifiers, linguistic qualifiers, etc.) it has been possible to attain a high human consistency and human friendliness. Many approaches have been proposed in this context, and most notable are the SQLf by Bosc and Pivert [5], and the introduction of fuzzy linguistic quantifiers first by Kacprzyk, Zadrożny and Ziółkowski [21, 20], which was finally extended and implemented as FQUERY for Access by Kacprzyk and Zadrożny [17, 18]. For a comprehensive survey we refer the reader to Zadrożny, De Tré, De Caluwe and Kacprzyk [28]. For a more basic and general account, Dubois and Prade [9] can also be useful.

Those approaches have provided querying tools with a considerable new potential by allowing for a direct use of natural language in queries. However, there are many more aspects of human specific elements that might be useful in database queries to make them even more human consistent and flexible. Notably, this concerns a natural human propensity to put preferences on querying conditions, even prioritize such conditions. The simplest, and presumably most important example is that some conditions may be considered mandatory, i.e. they must necessarily be fulfilled, and some conditions may be considered optional, i.e. they should be fulfilled if possible.

These conditions related to user preferences may be viewed in the perspective of *bipolarity* of information, as discussed in details by Dubois and Prade [11, 13, 1, 14], who have proposed a possibilistic framework for the representation and processing of such information.

Basically, a positive and negative information is concerned, and the negative information is related to mandatory conditions because it is meant as what cannot (is not allowed to) occur (hence the opposite, or complement, is to occur), while the positive information is related to what is possible, i.e. what can occur, so that it can be used to represent optional requirements, those which should be satisfied if possible.

This general framework is employed in this paper, both in the sense of its philosophy and some possibilistic tools. We are interested in the bipolarity of requirements (preferences) of the user while searching for information in a database.

Essentially, a database query may be identified with a *condition* (simple or complex, involving atomic conditions combined using some logical connectives) on the data sought that should be satisfied. As already mentioned, these conditions can be softened, and many approaches have been proposed in a broadly perceived area of flexible querying. First, *fuzzy predicates* for the modelling of *linguistic terms* in conditions has been advocated (cf., e.g., Zadrożny. De Treé, De Caluwe and Kacprzyk [28], Bosc and Pivert [5], Kacprzyk and Zadrożny [19]), the assignment of *importance weights* to particular parts of the condition has been proposed (cf., e.g., Dubois and Prade [9]), etc. For our purposes, bipolarity of information has also been allowed (cf., e.g., Dubois and Prade [10]) meant as that the user has in mind in fact two types of conditions:

- mandatory, or hard, constraints which have to be satisfied by the data sought, and
- optional constraints, or just preferences, making it possible to differentiate among the data items meeting the above mentioned mandatory constraints.

Such conditions, in the crisp case, imply clearly two sets of data items:

- *rejected*, *infeasible* etc., or, equivalently, taking a complement, *acceptable*, *satisfactory*, *feasible* etc., and
- *preferred*, *desired* etc.

The former conditions provide therefore the *negative* information indicating what should be avoided, while the latter provide the *positive* information indicating what is just preferred, hence the term "bipolar" to characterize queries comprising both types of conditions. The bipolar queries meant as above can be viewed from different perspectives and we will discuss this in Section 2. Moreover, we will discuss how to handle (combine) such two types of conditions when they are fuzzy, i.e. are satisfied *to a degree*.

This paper is mainly concerned with how to combine satisfaction degrees of both negative and positive conditions. The combination of the mandatory and possible conditions boils down to the definition (or selection from some available ones) of an appropriate aggregation operator. In the nonfuzzy context, the first attempt to introduce such type of bipolar queries was Lacroix and Lavency [22] followed by the development of a more general concept of a *query with preferences*, notably via its corresponding new relational algebra operator, called *winnow*, by Chomicki [6, 7].

These two approaches by Lacroix and Lavency and Chomicki, as well as some other related approaches which will not be discussed here, involve crisp (nonfuzzy) conditions only. Zadrożny [27] was first to propose a relatively straightforward, but effective and efficient, fuzzification of the approach by Lacroix and Lavency, and a further analysis of properties was given by Zadrożny and Kacprzyk [30]. A similar approach to the fuzzification of Chomicki's winnow operator was shown in Zadrożny and Kacprzyk [29]. The purpose of this paper is to present and account of those works, and present some new views and elements proposed by the authors.

2 Remarks on Bipolar Queries

Presumably, the term *bipolar queries* was first coined by Dubois and Prade [10]. However, for the purposes of this paper, it is expedient to start with an earlier approach which triggered a wider interest in this type of queries in the database community.

Lacroix and Lavency [22] were the first to propose a query involving two conditions: a *mandatory* condition (C) to be necessarily satisfied, and an *optional* condition that expresses just preferences (desires) (P), i.e. meant to be satisfied if possible. From a bipolar perspective, C may be seen to constitute a *negative* information (expressing the *negative* preferences): the tuples which do not satisfy it are definitely not matching the whole query. P, on the other hand, may be viewed to constitute a *positive* information (expressing the *positive preferences*): a tuple satisfying it is preferred over another tuple not satisfying it, provided both tuples satisfy the mandatory condition C.

We will identify the negative and positive conditions of a bipolar query with the predicates that represent them and also denote them as C and P, respectively. For a tuple $t \in T$, where $T = \{t_j\}$ denotes a set of tuples of a relation representing the database in question, $C(t)$ and $P(t)$ will denote that tuple t satisfies the respective condition. Then a bipolar query may be expressed in natural language as follows:

"Find tuples t satisfying (necessarily) C *and possibly* P"

The bipolar queries may be exemplified by:

"Find a house cheaper than 250 000 USD *and possibly* located not more than two blocks from a railway station" (1)

Here the negative condition excludes houses more expensive than 250 000 USD and the positive condition favors houses located closer to a railway station. Such a query may be more formally written as

$$C \text{ and possibly } P$$

or, equivalently, an answer to a bipolar query may be defined as the following set of tuples:

$$\{t : C(t) \textit{ and possibly } P(t)\} \tag{2}$$

and one can clearly see that a proper modeling of the aggregation of both types of query conditions, the mandatory and optional, which is expressed here with the use of the "and possibly" operator, is crucial. For some attempts to devise such an aggregation operator, see Yager [25, 24] and Bordogna and Pasi [2].

Obviously, from the perspective of the original (crisp) approach by Lacroix and Lavency, for such an operator the aggregation result depends not only on the arguments, $C(t)$ and $P(t)$, but also on the content of the database, in the following sense. If there are no tuples meeting both conditions, then the result of the aggregation is determined by the negative condition C alone. Otherwise, the aggregation boils down to a regular conjunction of both the conditions. This may best be expressed by the following logical formula [22]:

$$C(t) \text{ and possibly } P(t) \equiv C(t) \wedge \exists s(C(s) \wedge P(s)) \Rightarrow P(t) \tag{3}$$

This important property is therefore preserved under the "first select using C then order using P" rule that is behind the essence of the crisp bipolar query, i.e., the answer to the crisp bipolar query (C,P) is generated as follows:

- find tuples satisfying C,
- order them according to their satisfaction degree of condition P.

This view is predominant in fuzzy extensions of the original concept of Lacroix and Lavency. Both the direct extensions proposed by Bosc and Pivert [3, 4] as well as more sophisticated possibility theory based interpretation of this concept by Dubois and Prade [12] focus, in fact, on a proper treatment of *multiple* mandatory (required) and optional (preferred) conditions, basically assuming the above choice rule as the way of combining the negative and positive information.

To provide a point of departure for our discussion, let us briefly recall the approach to an aggregation of multiple positive conditions proposed for the crisp case by Lacroix and Lavency [22]. They consider the case where there is a set $\{P_i\}$ of preferred (positive) conditions which may be formally written as

$$C \text{ and possibly } \{P_i\} \tag{4}$$

The conditions P_i are meant to be combined in a non-standard way, i.e., they are not treated as a Boolean combination. Notably, two ways of their aggregation were proposed using aggregation operators based on the cardinality of the set of conditions P_i, and based on a varying importance of these conditions.

In the former case, a tuple satisfies query (4) if:

- it satisfies the required condition C, and
- there is no tuple s satisfying C and which satisfies more conditions P_i than tuple t satisfies.

In the latter case the positive conditions are assumed to be linearly ordered and tuple t satisfies query (4) if:

- it satisfies the required condition C, and
- there is no tuple s satisfying C and a condition P_i, while $\neg P_i(t)$ and $P_j(t) \equiv P_j(s)$ for all $j < i$.

For both types of such compound positive conditions an equivalent query in the relational calculus is defined, in the spirit of (3).

Bosc and Pivert [3] discuss some fuzzy counterparts of such types of compound positive conditions. For the cardinality based combination they consider a fuzzy set H_t of positive conditions P_i satisfied by a given tuple t, $\mu_{H_t}(P_i) = P_i(t) \in [0.1]$; notice that P_i is now a fuzzy condition, $P_i(t) \in [0,1]$ denotes its satisfaction degree by tuple t and a tuple satisfies (matches) the whole bipolar query to *a degree*. Then, the scalar cardinality (the so-called ΣCount) of H_t is used as the matching degree of tuple t with respect to the combination of the (normalized) positive conditions P_i. Bosc and Pivert [4] propose also a fuzzy counterpart of the importance based combination of the positive conditions by introducing a *hierarchical combination operator*, and the rule 'first select using C then order using P' is also followed.

Dubois and Prade in [10] define a bipolar query as a set of pairs (C_i, P_i) of negative and positive conditions,respectively, imposed on values of selected attributes $\{A_i\}_{i=1,k}$. These conditions may be identified with fuzzy sets defined in the domain of given attributes. These pairs of conditions are then combined (aggregated) to yield the overall conditions C and P as follows:

$$(C,P) = (\times_i C_i, +_i P_i), \tag{5}$$

where $\times_i C_i = C_1 \times C_2 \times \ldots \times C_k$, $+_i P_i = (P_1^c \times P_2^c \times \ldots \times P_k^c)^c$ and X^c is the complement of X.

Thus, the overall negative condition is obtained via the conjunction of all negative conditions concerning the particular attributes while the overall positive condition is obtained via the disjunction of all positive conditions concerning the particular attributes. Therefore, for the pair of overall conditions, we have

$$(C(t), P(t)) = (\min_i C_i(t), \max_i P_i(t)) \tag{6}$$

Then, the rule 'first select using C then order using P' is also followed which is done via the lexicographic order $\preceq$ of the tuples against the bipolar query, (5), that is:

$$t_1 \preceq t_2 \iff (C(t_1) < C(t_2)) \vee ((C(t_1) = C(t_2)) \wedge (P(t_1) \leq P(t_2))) \tag{7}$$

Dubois and Prade [10] consider also some non traditional (non-Boolean) combinations of the set of positive conditions P_i. Each tuple t is represented by a vector: $(C(t), P_{\sigma(1)}(t), \ldots, P_{\sigma(n)}(t))$ where σ is a permutation of the positive conditions P_i such that $P_{\sigma(1)}(t) \geq \ldots \geq P_{\sigma(n)}(t)$. Then, the lexicographic order of these vectors is used to rank order the tuples via the *leximax* operator (cf., e.g., [8]).

Recently, Dubois and Prade [13, 1, 12] presented a formal, possibility theory based framework for dealing with bipolar queries. Namely, two possibility distributions π and δ are assumed to represent query conditions (the user's preferences). The former corresponds to the negative (mandatory) condition, i.e., $\pi(t) = 1$ and $\pi(t) = 0$ mean, respectively, that tuple t is totally acceptable and totally unacceptable, with the intermediate values of $\pi(t)$ standing for an intermediate degree of acceptability. The latter possibility distribution δ represents the positive (optional) condition: $\delta(t) = 1$ denotes the maximum degree of preference (desirability) of t but $\delta(t) = 0$ means merely that t is not specifically preferred.

The above discussion concerns basically some fuzzy or possibilistic extensions of Lacroix and Lavency's original ideas. One can also view bipolar queries as a special case of queries which employ the well known concept of a *non-dominance relation* as recently proposed, for the crisp case, by Chomicki [6], and termed *queries with preferences*. A new relational algebra operator, called *winnow*, is introduced, which selects from a set of tuples those which are *non-dominated* with respect to a given *preference relation*, a binary relation on the set of tuples.

A bipolar query may then be obtained using a proper combination of the *select* operator with the winnow operator. The negative conditions define the select operator while the positive conditions are expressed by the preference relation. Therefore, the winnow operator may be easily combined with the traditional relational algebra operators. As in the case of queries proposed by Lacroix and Lavency [22], this combination may also be viewed to follow the 'first select then order' rule in the crisp case. This is not straightforward in the fuzzy case and some effective and efficient approach will be shown in the next section.

3 An Approach to Bipolar Queries with Fuzzy Conditions

We will first consider the Lacroix and Lavency [22] original approach to bipolar queries and extend it to the case of fuzzy conditions. Moreover, we will also propose a fuzzy version of the winnow operator and show its relation to "fuzzy" bipolar queries.

We start with the concept of a bipolar query exemplified by (1), and formalized by (2) and (3). Usually, the user will prefer to express the conditions in (1) using fuzzy predicates:

"Find a *cheap* house *and possibly* located *near* a railway station (8)

to be meant as that we are looking for a house that:

- has to be *cheap*,
- if there is a cheap house near the railway station then other, just cheap houses are of no (or maybe of a lesser) interest.

Notice that now the rule "first select using negative condition (here: cheap) then order using the positive condition (here: near the station)" cannot be directly applied as the properties involved are to a degree (from 0 to 1). For example, suppose that there is a house $H1$ definitely cheap (to the degree 1) but rather away from the station (near to the degree 0.2), and another house $H2$, still cheap but not that much as house $H1$ (for instance, to the degree 0.9) but located quite close to the station (to the degree 0.9). Which of them should belong to the answer to query (8)? By following "first select then order" rule, for house $H1$ we have a vector of satisfaction degrees $[1.0, 0.2]$ and for $H2$ a vector $[0.9, 0.9]$ so that the lexicographic order indicates that $H1$ is better than $H2$, which may evidently be questionable.

Now, we start with (3) and interpret it in terms of fuzzy logic. First, we will rewrite (3) using standard fuzzy counterparts of the logical connectives involved. Moreover, we will express it as the membership function of the resulting fuzzy set $ans(C,P,T)$ of tuples constituting the answer to the bipolar query (C,P) against a set of tuples T as:

$$\mu_{ans(C,P,T)}(t) = \min(C(t), \max(1 - \max_{s \in T} \min(C(s), P(s)), P(t))) \tag{9}$$

and T indicates that the membership degree (matching degree) of tuple t depends not only on this tuple itself and on the conditions C and P but also on the whole set of tuples T.

The *matching degree* of a tuple against a bipolar query is meant as the truth value of formula (3). Thus, the evaluation of a bipolar query produces a fuzzy set of tuples in which the membership function value for tuple t corresponds to the matching degree of this tuple against the query. The answer to a bipolar query is then a list of tuples, non-increasingly ordered according to their membership degree.

In (9) the min, max and $1-x$ operators are used to model the connectives of conjunction, disjunction and negation, respectively. Moreover, the implication $\Rightarrow$ is assumed to be the Kleene-Dienes implication (cf., e.g., [15] for a justification) and the existential quantifier $\exists$ is modeled via the maximum operator.

The formula (9) has been proposed by Yager [25, 24, 26] for an aggregation operator in the context of multicriteria decision making for so-called *possibilistically qualified criteria* intuitively characterized as those which should be satisfied unless they interfere with the satisfaction of other criteria. This is in fact the very essence of bipolar queries as considered in this paper.

A practically analogous concept was also applied by Bordogna and Pasi [2] in information retrieval. Moreover, Dubois and Prade [10] considered later a similar formula too. However, for (9) they employed an arbitrary parameter (instead of $\max_{s \in T} \min(C(s), P(s))$ in (9)) which implies that results obtained for a certain specific range of values $(C(t), P(t))$ may be difficult to justify. In Zadrożny's [27] proposal, which is the crucial element of the approach presented in this paper, this expression has a meaningful interpretation providing some justification for such a behavior.

Formula (9) is definitely just one of possible ways to fuzzify the original formula (3) proposed by Lacroix and Lavency [22] as different interpretations of the conjunction, disjunction and implication connectives may be employed, notably using various t-norms, t-conorms and implication operators (cf., e.g., [15]).

In particular one may consider so-called De Morgan Triples $(\wedge, \vee, \neg)$ that comprise a t-norm operator $\wedge$, a t-conorm operator $\vee$ and a negation operator $\neg$, such that $\neg(x \vee y) = \neg x \wedge \neg y$ holds.

The following three De Morgan Triples play the most important role in fuzzy logic (cf., e.g., Fodor and Roubens [15]):

$$(\wedge_{min}, \vee_{max}, \neg)$$
$$(\wedge_{\Pi}, \vee_{\Pi}, \neg)$$
$$(\wedge_{W}, \vee_{W}, \neg)$$

where:

$$
\begin{array}{lll}
x \wedge_{min} y & = \min(x,y) & \textit{minimum} \\
x \wedge_{\Pi} y & = x \cdot y & \textit{product} \\
x \wedge_{W} y & = \max(0, x+y-1) & \textit{Łukasiewicz t-norm} \\
x \vee_{max} y & = \max(x,y) & \textit{maximum} \\
x \vee_{\Pi} y & = x+y-x \cdot y & \textit{probabilistic sum} \\
x \vee_{W} y & = \min(1, x+y) & \textit{Łukasiewicz t-conorm}
\end{array}
$$

The negation operator $\neg$ in all the cases is defined as: $\neg x = 1 - x$.

In fuzzy logic the general and existential quantifiers are equated, for the case of a finite universe, with the maximum and minimum operators, respectively, in the sense that:

$$
\begin{aligned}
\text{truth}(\forall x A(x)) &= \min_x \mu_A(x) \\
\text{truth}(\exists x A(x)) &= \max_x \mu_A(x).
\end{aligned}
$$

Thus we adopt the following definitions:

$$\text{truth}(\forall x A(x)) = \mu_A(a_1) \wedge \mu_A(a_2) \wedge \ldots \wedge \mu_A(a_m) \quad (10)$$

$$\text{truth}(\exists x A(x)) = \mu_A(a_1) \vee \mu_A(a_2) \vee \ldots \vee \mu_A(a_m) \quad (11)$$

There are two most popular ways of deriving an implication operator with respect to a given De Morgan Triple $(\wedge, \vee, \neg)$, namely so-called S-implications and R-implications defined as follows:

$$R-\text{implication:} \quad x \rightarrow y = \sup\{z : x \wedge z \leq y\} \quad (12)$$

$$S-\text{implication:} \quad x \rightarrow y = \neg x \vee y \quad (13)$$

Thus, for the particular De Morgan Triples one obtains the following R-implication operators:

Gödel's implication $\quad x \rightarrow_{R-min} y = \begin{cases} 1 & \text{for } x \leq y \\ y & \text{for } x > y \end{cases}$

Goguen's implication $\quad x \rightarrow_{R-\Pi} y = \begin{cases} 1 & \text{for } x = 0 \\ \min\{1, \frac{y}{x}\} & \text{for } x \neq 0 \end{cases}$

Łukasiewicz' implication $\quad x \rightarrow_{R-W} y = \min(1 - x + y, 1)$

and the following S-implication operators:

Kleene–Dienes' implication $\quad x \rightarrow_{S-max} y = \max(1-x, y)$

Reichenbach's implication $\quad x \rightarrow_{S-\Pi} y = 1 - x + x \cdot y$

The S-implication operator $\rightarrow_{S-W}$ is identical with $\rightarrow_{R-W}$.

In order to simplify the notation let us fix C, P and T in (9) and denote its version for a given De Morgan Triple, its related R or S implication and a corresponding existential quantifier as, respectively, $\gamma_{\wedge,R}$ and $\gamma_{\wedge,S}$. Thus, for example $\gamma_{\min,S}(t) = \mu_{ans(C,P,T)}(t)$ denotes the original version of (9).

In Table 1 various emerging interpretations of (9) are shown.

Table 1. Right-hand side of the formula (9) for different interpretations of the logical connectives

$\gamma_{\wedge,\cdot}$	**Resulting form of the formula (9)**
$\gamma_{min,S}$	$\min(C(t), \max(1 - \max_{s \in T} \min(C(s), P(s)), P(t)))$
$\gamma_{min,R}$	$\begin{cases} C(t) & \text{if } \max_{s \in T} \min(C(s), P(s)) \leq P(t) \\ \min(C(t), P(t)) & \text{otherwise} \end{cases}$
$\gamma_{\Pi,S}$	$C(x) \cdot (\prod_i (1 - C(y_i) \cdot P(y_i)) \cdot (1 - P(x)) + P(x))$
$\gamma_{\Pi,R}$	$\begin{cases} C(t) & \text{if } \exists_{\Pi}(C(s_i) \cdot P(s_i)) = 0 \\ C(t) \cdot \min(\frac{P(t)}{\exists_{\Pi}(C(s_i) \cdot P(s_i))}, 1) & \text{otherwise} \end{cases}$
γ_W	$C(t) \wedge_W (\exists_W (C(s) \wedge_W P(s)) \rightarrow_W P(t))$

A proper choice of the logical connectives (including the existential quantifier) in (3) may be done in two ways. First, one may look for some properties and try to check which operators modeling the connectives, to be called *logical operators*, provide for the expected behavior. Second, one can study the properties under different logical operators. We propose some results of the research along both lines in [30, 31], which may be summarized as follows.

Basically, for any choice of the logical operators a characteristic feature of bipolar queries is preserved: if there is a tuple satisfying both the mandatory (required) and optional (preferred) conditions, then the combination of them is via the conjunction. On the other hand if for a tuple $t \in T$, $P(t) = 1$, then the result is $C(t)$ which is implied by: $x \rightarrow 1 = 1$ and $x \wedge 1 = x$, for any $\wedge$ and $\rightarrow$ (cf., e.g., [15]). Thus, if a tuple fully satisfies the positive condition P, then its overall matching degree is equal to its satisfaction of the negative condition C.

But, even more important question is: does the choice of logical operators influence the resulting order of tuples in the answer to a bipolar query ? Basically, this is the case and we refer the reader for details to Zadrożny and Kacprzyk [30]. Moreover, in general, the choice between an S-implication and an R-implication, keeping all other logical operators fixed, may change the order of the tuples, but not necessarily.

4 Queries with Preferences and Bipolar Queries

We will briefly discuss now the concept of a query with preferences, introduced by Chomicki [6, 7], which may be conveniently presented in terms of a new operator of relational algebra called *winnow*. This is an unary operator which selects from a set of tuples those which are *non-dominated* with respect to a given preference relation. Chomicki [6, 7] defines this operator for the crisp case only, i.e., for crisp (nofuzzy) preference relations and sets of tuples. We propose a fuzzy version of the *winnow* operator and show its relation to bipolar queries.

The *winnow* operator is defined with respect to a *preference relation* which is any binary relation R defined on the set of tuples T, $R \subseteq T \times T$.

First, if two tuples $t, s \in T$ are in relation R, i.e., $R(t,s)$, then it is said that tuple t *dominates* tuple s with respect to relation R.

Let T be a set of tuples and R a preference relation defined on T. Then the *winnow* operator ω_R, $\omega_R : T \longrightarrow 2^T$, is defined as

$$\omega_R(T) = \{t \in T : \neg\exists_{s \in T}\ R(s,t)\} \tag{14}$$

Thus, for a given set of tuples T it yields a subset of the *non-dominated* tuples with respect to R.

A relational algebra query employing the *winnow* operator is referred to as a *query with preferences*. It may be easily shown (cf. Chomicki [6]) that the *winnow* operator may be expressed as a combination of the standard classical relational algebra operators but it is better to consider it as a distinguished operator to easier study its properties and behavior.

The concept of a *winnow* operator may be illustrated on the following simple example. Consider a database of a real-estate agency with a table `HOUSES` describing the details of particular real-estate properties offered by the agency (each house is represented by a tuple). The schema of the relation `HOUSES` contains, among possibly many other ones, the attributes `city` and `price`. Assume that we are interested in the list of the *cheapest* houses in each city. Then the preference relation should be defined as follows

$$R(t,s) \Leftrightarrow (t.\texttt{city} = s.\texttt{city}) \wedge (t.\texttt{price} < s.\texttt{price})$$

where $t.A$ denotes the value of attribute A (e.g., `price`) in tuple t. Therefore, the *winnow* operator $\omega_R(\texttt{HOUSES})$ will select the houses that are sought (here a database table, such as `HOUSES`, is treated as a set of tuples). Indeed, according to the definition of the *winnow* operator, we will get as an answer a set of houses, which are non-dominated with respect to R, i.e., for which there is no other house in the same city which has a lower price.

A bipolar query with the *crisp* conditions: the negative C and the positive P may be expressed using the *winnow* operator as follows, employing the example (1) of a bipolar query.

The preference relation R should be defined as:

$$R(t,s) \Leftrightarrow (t.\texttt{to\_station} \leq 2) \wedge (s.\texttt{to\_station} > 2)$$

assuming that `to_station` indicates how many blocks away is the house from the closest railway station. Then, the following relational algebra query with a *winnow* operator yields the required results

$$\omega_R(\sigma_{\texttt{price} \leq 250000}(\texttt{HOUSES}))$$

where σ_ϕ is the classical *selection* operator that selects from a set of tuples those for which ϕ holds. This query preserves the characteristic property of bipolar queries as discussed earlier, i.e., if there are houses cheaper than 250 000 USD and located closer than two blocks from the station, then only they will be selected (the houses satisfying only the negative condition will be ignored). Otherwise, all houses satisfying the negative condition will be selected, if they exist.

A general scheme for translating a bipolar query characterized by a pair of negative and positive conditions, (C,P), to a corresponding query with preferences is therefore: the preference relation R is defined as

$$R(t,s) \Leftrightarrow P(t) \wedge \neg P(s) \tag{15}$$

and then the overall query with preferences is:

$$\omega_R(\sigma_C(T))$$

Now we will propose a fuzzy counterpart of the *winnow* operator which also will make it possible to express (fuzzy) bipolar queries. We have to take into account that:

- R is a *fuzzy preference relation*,
- a fuzzy counterpart of *non-dominance* has to be employed,
- the set of tuples T is a *fuzzy set*.

It is convenient to use the concept of a *fuzzy choice function* (cf. Świtalski [23]) since then the set of non-dominated elements with respect to a fuzzy preference relation may be conveniently expressed. Let us start with a concept of a crisp set $R^-(s)$, defined as:

$$R^-(s) = \{u \in T : R(s,u)\} \tag{16}$$

and gathering all tuples dominated by a tuple s with respect to a crisp preference relation R. Then, $N(T,R)$, defined as follows:

$$N(T,R) = T \cap \bigcap_{s \in T} \overline{R^-(s)} \tag{17}$$

denotes the set of all non-dominated tuples of a (crisp) set of tuples T with respect to a (crisp) preference relation R, while $\overline{A}$ denotes the complement of A. For a further fuzzification it is convenient to rewrite (17) as a predicate calculus formula

$$N(T,R)(t) \Leftrightarrow T(t) \wedge \forall_{s \in T} \neg R^-(s)(t) \tag{18}$$

where the particular predicates are denoted with the same symbols as their corresponding sets (in particular, $R^-(s)$ denotes a predicate corresponding to set (16) defined for tuple s).

Using (17) we may define the *winnow* operator, equivalent to (14), as:

$$\omega_R(T) = N(T,R) \tag{19}$$

Now, let us adapt (19) to the case of a fuzzy preference relation R on a crisp set of tuples T, characterized by its membership function $\mu_{\tilde{R}}$. The dominance (and non-dominance) naturally becomes now a matter of degree. Thus we define a fuzzy set of tuples that are non-dominated with respect to a fuzzy preference relation $\tilde{R}$, using (16)–(17) and interpreting the set operations of the intersection and complement as the standard operations on fuzzy sets. We start with a fuzzy counterpart of the set (16), defining the membership function of the fuzzy set $\tilde{R}^{-}(s)$ of tuples dominated (to a degree) by tuple s with respect to the fuzzy preference relation $\tilde{R}$:

$$\mu_{\tilde{R}^{-}(s)}(u) = \mu_{\tilde{R}}(s,u) \tag{20}$$

Next let us rewrite (18), replacing a preference relation R with a fuzzy preference relation $\tilde{R}$ and replacing R^{-} with $\tilde{R}^{-}$, according to (20):

$$N(T,\tilde{R})(t) \Leftrightarrow T(t) \wedge \forall_{s \in T} \neg \tilde{R}(s,t) \tag{21}$$

We still have to take into account that set T (and a predicate corresponding to it) is, in general, fuzzy. Thus we denote it as $\tilde{T}$ and replace the restricted quantifier $\forall_{s \in T}$ in (21) with an equivalent non-restricted form obtaining:

$$N(\tilde{T},\tilde{R})(t) \Leftrightarrow \tilde{T}(t) \wedge \forall_{s} \, (\tilde{T}(s) \rightarrow \neg \tilde{R}(s,t)) \tag{22}$$

Finally, we can define a fuzzy counterpart of the *winnow* operator in the following way. Let $\tilde{T}$ be a fuzzy set of tuples and $\tilde{R}$ be a fuzzy preference relation, both defined on the same set of tuples T. Then the *fuzzy winnow operator* $\omega_{\tilde{R}}$ is defined as:

$$\omega_{\tilde{R}}(\tilde{T})(t) = N(\tilde{T},\tilde{R})(t) \qquad t \in T \tag{23}$$

where the fuzzy predicate $N(\tilde{T},\tilde{R})$ is determined by (22), and $\omega_{\tilde{R}}(\tilde{T})(t)$ denotes the value of the fuzzy membership function of the set of tuples defined by $\omega_{\tilde{R}}(\tilde{T})$ for a tuple t.

Again, as in case of the fuzzy bipolar queries, one may study the effect of the choice of various logical operators to model logical connectives in (22) but this will not be considered here. Now we will just show how a bipolar query may be expressed using the concept of the fuzzy *winnow* operator.

Let us consider a bipolar query defined by a pair of fuzzy conditions (C,P). These conditions will be identified with fuzzy predicates, denoted with the same symbols, for simplicity. Let $\tilde{R}$ be a fuzzy preference relation given as (cf., (15)):

$$\tilde{R}(t,s) \Leftrightarrow P(t) \wedge \neg P(s) \tag{24}$$

Then the bipolar query may be expressed as the following combination of the selection and fuzzy *winnow* operators:

$$\omega_{\tilde{R}}(\sigma_C(T)) = N(C(T),\tilde{R}) \tag{25}$$

where $C(T)$ is a fuzzy set of the elements of T satisfying (to a degree) the condition C, i.e., $\mu_{C(T)}(t) = C(t)$.

Using (22) we can define the predicate (set) $N(C(T),\tilde{R})$ in (25) as follows:

$$N(C(T),\tilde{R})(t) \Leftrightarrow C(t) \wedge \forall_s \, (C(s) \rightarrow \neg(P(s) \wedge \neg P(t))) \tag{26}$$

Note that the selection operator σ_C in (25) may also be applied to a fuzzy set of tuples T, what may be convenient if the set of tuples T is a result of another fuzzy query.

In Zadrożny and Kacprzyk [29] we show that for the conjunction, negation and implication connectives in (26) modeled by the minimum, $n(x) = 1 - x$ and the Kleene-Dienes implication, respectively, the fuzzy set of tuples obtained using (25) is identical with the fuzzy set defined by (9).

5 Concluding Remarks

We briefly discussed an approach to bipolar queries. We start with the original idea proposed in Lacroix and Lavency [22] and briefly review some selected relevant approaches recently proposed in the literature. In particular we point out two main lines of research, the one focusing on a formal representation within some well established theories and the analysis of a meaningful combinations of multiple conditions, and the one concerned mainly with the study of how to aggregate mandatory (negative, required) and optional (positive, or desired) conditions. We follow the second line of research and show some relations with other approaches, both concerning database querying (exemplified by Chomicki [6]) as well as other domains, mainly in multicriteria decision making (exemplified by Yager [25]). In the former case we offer a fuzzy counterpart of a new relational algebra operator *winnow* .

References

1. Benferhat, S., Dubois, D., Kaci, S., Prade, H.: Modeling positive and negative information in possibility theory. International Journal of Intelligent Systems 23(10), 1094–1118 (2008)
2. Bordogna, G., Pasi, G.: Linguistic aggregation operators of selection criteria in fuzzy information retrieval. International Journal of Intelligent Systems 10(2), 233–248 (1995)
3. Bosc, P., Pivert, O.: Discriminated answers and databases: fuzzy sets as a unifying expression means. In: Proceedings of the IEEE International Conference on Fuzzy Systems (FUZZ-IEEE), San Diego, USA, pp. 745–752 (1992)
4. Bosc, P., Pivert, O.: An approach for a hierarchical aggregation of fuzzy predicates. In: Proceedings of the Second IEEE International Conference on Fuzzy Systems (FUZZ-IEEE 1993), San Francisco, USA, pp. 1231–1236 (1993)
5. Bosc, P., Pivert, O.: SQLf: A relational database language for fuzzy querying. IEEE Transactions on Fuzzy Systems 3(1), 1–17 (1995)
6. Chomicki, J.: Querying with intrinsic preferences. In: Jensen, C.S., Jeffery, K., Pokorný, J., Šaltenis, S., Bertino, E., Böhm, K., Jarke, M. (eds.) EDBT 2002. LNCS, vol. 2287, pp. 34–51. Springer, Heidelberg (2002)

7. Chomicki, J.: Preference formulas in relational queries. ACM Transactions on Database Systems 28(4), 427–466 (2003)
8. Dubois, D., Fargier, H., Prade, H.: Refinement of the maximin approach to decision-making in fuzzy environment. Fuzzy Sets and Systems (81), 103–122 (1996)
9. Dubois, D., Prade, H.: Using fuzzy sets in flexible querying: why and how? In: Andreasen, T., Christiansen, H., Larsen, H. (eds.) Flexible Query Answering Systems, pp. 45–60. Kluwer Academic Publishers, Dordrecht (1997)
10. Dubois, D., Prade, H.: Bipolarity in flexible querying. In: Andreasen, T., Motro, A., Christiansen, H., Larsen, H. (eds.) FQAS 2002. LNCS (LNAI), vol. 2522, pp. 174–182. Springer, Heidelberg (2002)
11. Dubois, D., Prade, H.: Bipolar representations in reasoning, knowledge extraction and decision processes. In: Greco, S., Hata, Y., Hirano, S., Inuiguchi, M., Miyamoto, S., Nguyen, H.S., Słowiński, R. (eds.) RSCTC 2006. LNCS (LNAI), vol. 4259, pp. 15–26. Springer, Heidelberg (2006)
12. Dubois, D., Prade, H.: Handling bipolar queries in fuzzy information processing. In: Galindo [16], pp. 97–114
13. Dubois, D., Prade, H.: An introduction to bipolar representations of information and preference. International Journal of Intelligent Systems 23(8), 866–877 (2008)
14. Dubois, D., Prade, H.: An overview of the asymmetric bipolar representation of positive and negative information in possibility theory. Fuzzy Sets and Systems 160(10), 1355–1366 (2009)
15. Fodor, J., Roubens, M.: Fuzzy Preference Modelling and Multicriteria Decision Support. Series D: System Theory, Knowledge Engineering and Problem Solving. Kluwer Academic Publishers, Dordrecht (1994)
16. Galindo, J. (ed.): Handbook of Research on Fuzzy Information Processing in Databases. Information Science Reference, New York (2008)
17. Kacprzyk, J., Zadrożny, S.: Fuzzy querying for Microsoft Access. In: Proceedings of the Third IEEE Conference on Fuzzy Systems (FUZZ-IEEE 1994), Orlando, USA, vol. 1, pp. 167–171 (1994)
18. Kacprzyk, J., Zadrożny, S.: Fuzzy queries in Microsoft Access v. 2. In: Proceedings of 6th International Fuzzy Systems Association World Congress, Sao Paolo, Brazil, vol. II, pp. 341–344 (1995)
19. Kacprzyk, J., Zadrożny, S.: Computing with words in intelligent database querying: standalone and internet-based applications. Information Sciences 134(1-4), 71–109 (2001)
20. Kacprzyk, J., Zadrożny, S., Ziółkowski, A.: FQUERY III+: a "human consistent" database querying system based on fuzzy logic with linguistic quantifiers. Information Systems 14(6), 443–453 (1989)
21. Kacprzyk, J., Ziółkowski, A.: Database queries with fuzzy linguistic quantifiers. IEEE Transactions on System, Man and Cybernetics 16(3), 474–479 (1986)
22. Lacroix, M., Lavency, P.: Preferences: Putting more knowledge into queries. In: Proceedings of the 13th International Conference on Very Large Databases, Brighton, UK, pp. 217–225 (1987)
23. Świtalski, Z.: Choice functions associated with fuzzy preference relations. In: Kacprzyk, J., Roubens, M. (eds.) Non-Conventional Preference Relations in Decision Making, pp. 106–118. Springer, Berlin (1988)
24. Yager, R.: Fuzzy sets and approximate reasoning in decision and control. In: Proceedings of the IEEE International Conference on Fuzzy Systems (FUZZ-IEEE), San Diego, USA, pp. 415–428 (1992)
25. Yager, R.: Higher structures in multi-criteria decision making. International Journal of Man-Machine Studies 36, 553–570 (1992)

26. Yager, R.: Fuzzy logic in the formulation of decision functions from linguistic specifications. Kybernetes 25(4), 119–130 (1996)
27. Zadrożny, S.: Bipolar queries revisited. In: Torra, V., Narukawa, Y., Miyamoto, S. (eds.) MDAI 2005. LNCS (LNAI), vol. 3558, pp. 387–398. Springer, Heidelberg (2005)
28. Zadrożny, S., De Tre, G., De Caluwe, R., Kacprzyk, J.: An overview of fuzzy approaches to flexible database querying. In: Galindo [16], pp. 34–53
29. Zadrożny, S., Kacprzyk, J.: Bipolar queries and queries with preferences. In: Proceedings of the 17th International Conference on Database and Expert Systems Applications (DEXA 2006), pp. 415–419. IEEE Comp. Soc., Krakow (2006)
30. Zadrożny, S., Kacprzyk, J.: Bipolar queries using various interpretations of logical connectives. In: Melin, P., Castillo, O., Aguilar, L.T., Kacprzyk, J., Pedrycz, W. (eds.) IFSA 2007. LNCS (LNAI), vol. 4529, pp. 181–190. Springer, Heidelberg (2007)
31. Zadrożny, S., Kacprzyk, J.: Bipolar queries: an approach and its various interpretations. In: Proceedings of the IFSA Congress/EUSFLAT Conference, Lisbon, Portugal (to appear, 2009)

On Some Uses of a Stratified Divisor in an Ordinal Framework

Patrick Bosc and Olivier Pivert

Abstract. In this paper, we are interested in taking preferences into account for division-like queries. The interest for introducing preferences is first to cope with user needs, then to get discriminated results instead of a flat set of elements. Here, the idea is to use ordinal preferences which are not too demanding for a casual user. Moreover, the type of query considered is inspired by the division operator and some of its variations where preferences apply only to the divisor. The division aims at retrieving the elements associated with a specified set of values and in a similar spirit, the anti-division looks for elements which are associated with none of the values of a given set. One of the focuses of this paper is to investigate queries mixing those two aspects. In order to remain coherent with the denomination of (anti-)division, the property of the result delivered is characterized. Last, a special attention is paid to the implementation of such queries using a regular database management system and some experimental results illustrate the feasibility of the approach.

Keywords: Relational databases, division, anti-division, quotient, ordinal preferences.

1 Introduction

Queries including preferences have received a growing interest during the last decade [1, 2, 5, 6, 8, 11, 12, 13]. One of their main advantages is to allow for some discrimination among the elements of their result (which is no longer a flat set) thanks to the compliance with the specified preferences. However, up to now, most of the research works have focused on fairly simple queries where preferences apply only to selections. The objective of this paper is to enlarge the scope of queries concerned with preferences by considering more complex queries,

---

Patrick Bosc and Olivier Pivert
IRISA/ENSSAT - University of Rennes 1, 6 Rue de Kerampont
Technopole Anticipa BP 80518, 22305 Lannion Cedex, France

J. Kacprzyk et al. (Eds.): Uncer. Appro. for Spatial Data Model. and Process., SCI 271, pp. 133–154.
springerlink.com

founded on the association of an element with a given set of values, in the spirit of the division operation. Moreover, a purely ordinal framework is chosen and the user has only to deal with an ordinal scale, which we think to be not too demanding. Last, taking preferences into account will allow for keeping only the best k answers, in the spirit of top-k queries [5].

Knowing that a regular division delivers a non discriminated set of elements, the idea is to call on preferences related to the divisor. Two major lines for assigning preferences may be thought of, depending on whether they concern tuples individually (see e.g., [2]), or (sub)sets of tuples, which is the choice made here and we will use the term "stratified divisor".

Moreover, we will not only consider the division, but a neighbor operator called the anti-division. As the division retrieves elements associated with a given set of values, the anti-division looks for elements that are associated with none of the elements of a specified set.

In both cases, the first layer of the divisor may be seen as an initial divisor and the following layers serve to break ties between elements associated with all (respectively none of) the values of the first layer. In other words, to be satisfactory, an element x of the dividend must be associated with all (respectively none of) the values of the first layer. The way the next layers are taken into account is discussed in more details in the body of the paper.

Such extended division (respectively anti-division) queries can be expressed in natural language as:

> "find the elements x connected **in priority** with all (respectively none) of $\{set\}_1$ **then if possible** with all (respectively none) of $\{set\}_2$ ... **then if possible** with all (respectively none) of $\{set\}_n$".

This type of statement has some relationship with bipolarity [9, 10]. Indeed, this falls in the third category of bipolarity reported in [10] where the two types of criteria are of a different nature. Here, the connection with all (respectively none of) the values of $\{set\}_1$ is a constraint and those with all (respectively none of) the values of $\{set\}_2$ to $\{set\}_n$ represent wishes which are not mandatory (in the sense of acceptance/rejection).

Let us illustrate the idea of an extended division with a user looking for wine shops offering Saint Emilion Grand Cru, Pomerol and Margaux and if possible Gewurztraminer Vendanges Tardives and Chablis Premier Cru and if possible Pommard and Chambertin. Similarly, an anti-division is of interest if one is interested in food products which do not contain some additives, where some are totally forbidden and other more or less undesired.

The rest of the paper is organized as follows. Section 2 is devoted to some reminders about the division and anti-division operators in the usual relational setting. In section 3 a stratified version of these operators is presented along with their syntax and modeling. It is also shown that the result they deliver has the same property as in the usual case. In Section 4, the issue of considering queries involving both a stratified division and a stratified anti-division is tackled. Implementation issues for all these queries involving stratified operations are discussed in section 5. The conclusion summarizes the contributions of the paper and evokes some lines for future work.

2 The Regular Division and Anti-division Operators

In the rest of the paper, the dividend relation r has the schema (A, X), while without loss of generality that of the divisor relation s is (B) where A and B are compatible sets of attributes, i.e., defined on the same domains of values.

2.1 *The Division*

The relational division, i.e., the division of relation r by relation s is defined as:

$$\text{div}(r, s, A, B) = \{x \mid (x \in r[X]) \wedge (s \subseteq \Omega_r(x))\} \quad (1)$$

$$= \{x \mid (x \in R[X]) \wedge (\forall a, a \in s \Rightarrow (a, x) \in r)\} \quad (2)$$

where r[X] denotes the projection of r over X and $\Omega_r(x) = \{a \mid <a, x> \in r\}$. In other words, an element x belongs to the result of the division of r by s if and only if it is associated in r with **at least all** the values a appearing in s. The justification of the term "division" assigned to this operation relies on the fact that a property similar to that of the quotient of integers holds. Indeed, the resulting relation res obtained with expression (1) has the double characteristic of a quotient:

$$s \times res \subseteq r \quad (3a)$$

$$\forall res' \supset res, s \times res' \not\subseteq r \quad (3b)$$

× denoting the Cartesian product of relations. Expressions (3a) and (3b) express the fact that the relation res resulting from the division (according to formula (1) or (2)) is a quotient, i.e., **the largest relation** whose Cartesian product with the divisor returns a result smaller than or equal to the dividend (according to regular set inclusion).

In an SQL-like language, the division of r by s can be expressed thanks to a partitioning mechanism:

select X **from** r [**where** condition] **group by** X
having set(A) **contains** (**select** B **from** s **where** …).

Example 1. Let us take a database involving the two relations order (o) and product (p) with respective schemas O(np, store, qty) and P(np, price). Tuples <n, s, q> of o and <n, pr> of p state that the product whose number is n has been ordered from store s in quantity q and that its price is pr. Retrieving the stores which have been ordered all the products priced under $127 in a quantity greater than 35, can be expressed thanks to a division as:

div(o-g35, p-u127, {np}, {np})

where relation o-g35 corresponds to pairs (n, s) such that product n has been ordered from store s in a quantity over 35 and relation p-u127 gathers products whose price is under $127. From the following extensions of relations o-g35 and p-u127:

o-g35 = {<15, 32>, <12, 32>, <34, 32>, <26, 32>, <12, 7>, <26, 7>,
<15, 19>, <12, 19>, <26, 19>},
p-u127 = {<15>, <12>, <26>},

the previous division using formula (1) leads to a result made of two elements {<32>, <19>}. It can easily be checked that this result satisfies expressions (3a) and (3b). ♦

2.2 *The Anti-division*

Similarly, we call anti-division the operator ⋇ defined the following way:

$$r\,[A \divideontimes B]\; s = \{x \mid (x \in r[X]) \wedge (s \subseteq cp(\Omega_r(x)))\} \quad (4)$$

$$= \{x \mid (x \in r[X]) \wedge (\forall a,\, a \in s[B] \Rightarrow (a, x) \notin r)\} \quad (5)$$

where cp(rel) denotes the complement of rel. The result ad-res of the anti-division may be called an "anti-quotient", i.e., the largest relation whose Cartesian product with the divisor is included in the complement of the dividend. Thus, the following two properties hold:

$$s \times \text{ad-res} \subseteq cp(r) \quad (6a)$$

$$\forall \text{ad-res'} \supset \text{ad-res},\; s \times \text{ad-res'} \not\subseteq cp(r). \quad (6b)$$

In an SQL-like language, the anti-division of r by s can be expressed in a way similar to a division:

select X **from** r [**where** condition] **group by** X
having set(A) **contains-none** (**select** B **from** s **where** …)

where the operator "contains-none" states that the two operand sets do not overlap. An alternative expression is based on a difference:

(**select** X **from** r) **differ** (**select** X **from** r **where** A **in** (**select** B **from** s)).

Example 2. Let us consider the following relations Prod(product, component, proportion), which describes the composition of some chemical products and Nox(component) which gathers the identifications of noxious components:

Prod = $\{<p_1, c_1, 3>, <p_1, c_2, 4>, <p_1, c_3, 54>, <p_2, c_1, 9>, <p_2, c_4, 30>,$
$<p_3, c_2, 8>, p_3, c_6, 22>\}$,

Nox = $\{<c_1>, <c_2>, <c_5>\}$.

The query "retrieve any product which does not contain any noxious component in a proportion higher than 5%" can be expressed as the anti-division of the relation Prod1 derived from Prod (on the basis of a proportion over 5%) made of:

$$\{<p_1, c_3>, <p_2, c_1>, <p_2, c_4>, <p_3, c_2>, <p_3, c_6>\}$$

by Nox, whose result according to (4) or (5) is $\{p_1\}$ and it is easy to check that formulas (6a-6b) both hold. ♦

3 Stratified Division and Anti-division Queries

In this section, we first give some characteristics of the stratification. Then, the expression of stratified division and anti-division queries in an SQL-like fashion is proposed as well as the modelling of such queries. Finally, the property of the result delivered is discussed.

3.1 *About the Stratification Mechanism*

As mentioned before, the key idea is to use a divisor made of several layers. So, there is a preference relation over the subsets of the divisor, namely:

$$(S_1 = \{v_{1,1}, \ldots, v_{1,j_1}\}) \succ \ldots \succ (S_n = \{v_{n,1}, \ldots, v_{n,j_n}\})$$

where $a \succ b$ denotes the preference of a over b. Associated with this preference relation is an ordinal scale L with labels l_i's such that:

$$l_1 > \ldots > l_n > l_{n+1}$$

which will be also used to assign levels of satisfaction to elements pertaining to the result of any stratified division or anti-division. In this scale, l_1 is the maximal element for the highest satisfaction and the last label l_{n+1} expresses rejection. These two specific levels play the role of 1 and 0 in the unit interval.

Example 3. Coming back to the example of the query related to wine shops evoked in the introduction, there are three layers:

S_1 = {Saint Emilion Grand Cru, Pomerol, Margaux},

S_2 = {Gewurztraminer Vendanges Tardives, Chablis Premier Cru},

S_3 = {Pommard, Chambertin},

along with the scale $L = l_1 > l_2 > l_3 > l_4$. ♦

According to the view adopted in this paper, the first stratum S_1 is considered mandatory, whereas the next ones (S_2 to S_n) define only wishes. In other words, S_1 is a regular divisor and $S_2, \ldots, S_n$ are introduced as complementary components in order to discriminate among the elements of the dividend associated with all (respectively none) of the values of S_1. In addition, the layers are considered in a hierarchical fashion, which means that a given layer intervenes only if the association (or non-association) with all the previous ones holds. This behavior is similar to what is done in the systems Preferences [13] and PreferenceSQL [12] or with the operator winnow [6] when cascades of preferences are used. Finally, an element x of the dividend is all the more acceptable as it is (respectively it is not) connected with a "long" succession of layers of the divisor starting with S_1. In other words, x is preferred to y if x is associated with all (respectively none) of the values of the sets S_1 to S_p and y is associated with all (respectively none) of the elements of a shorter list of sets.

Example 4. Let us consider the stratified divisor :

s = {{a, b, c}, {d}, {e, f}}

and the dividend:

r = {<x1, a>, <x1, b>, <x1, c>, <x1, d>, <x1,e>,
<x2, a>, <x2, b>, <x2, c>,
<x3, a>, <x3, b>,<x3, c>, <x3, e>, <x3, f>,
<x4, a>, <x4, e>, <x4, f>,
<x5, b>, <x5, c>,
<x6, d>, <x6, e>}.

The stratified division of r by s discards x4, x5 and x6 which are not exhaustively associated with S_1 = {a, b, c} and it delivers the result: x1 $\succ$ {x2, x3}. ♦

It must be noticed that the view adopted here is somehow conjunctive. An alternative would be to model a behavior that takes into account all the layers in a hierarchical way and build, for a given x, a vector E(x) of Boolean values (E(x)[i] = 1 if x is associated with all (respectively none) of the values from layer S_i, 0 otherwise). The different x's could then be ranked according to the lexicographic order over the vectors.

3.2 *Syntax of Stratified Operations*

Division queries are expressed in an SQL-like style where the dividend may be any intermediate relation (not only a base relation) and the divisor is either explicitly given by the user, or stated thanks to subqueries, along with his/her preferences. This is done in way quite similar to the usual division (i.e., thanks to a partitioning mechanism), namely:

select top k X **from** r [**where** condition] **group by** X
having set(A) **contains** $\{v_{1,1}, \ldots, v_{1,j_1}\}$ **and if possible** …
and if possible $\{v_{n,1}, \ldots, v_{n,j_n}\}$.

Coming back to the example of wines evoked before, such a query could be:

select top 6 shop-name **from** wineshops **group by** shop-name
having set(wine) **contains** {Saint Emilion Grand Cru, Pomerol, Margaux}
and if possible {Gewurztraminer Vendanges Tardives, Chablis Premier Cru}
and if possible {Pommard, Chambertin}.

In the context of medical diagnosis, the following example illustrates the use of subqueries to build the stratified divisor. Let us consider: i) a relation disease(name, symptom, frequency) which describes the symptoms associated with some diseases as well as the frequency with which a given symptom appears for a given disease, ii) a relation patient(#person, symptom) which describes the symptoms shown by some patients. The following stratified division query looks for the patients which have all of the 100% frequent symptoms of influenza, and if possible all of the symptoms whose frequency is above 80%, and if possible all of the symptoms whose frequency is above 50%:

select top 10 #person **from** patient
group by #person
having set(symptom) **contains**
(**select** symptom **from** disease **where** name = 'flu' **and** frequency = 100)
and if possible
(**select** symptom **from** disease
where name = 'flu' **and** frequency **between** 80 **and** 99)
and if possible
(**select** symptom **from** disease
where name = 'flu' **and** frequency **between** 50 **and** 79)

The anti-division is similarly formulated as:

select top k X **from** r [**where** condition] **group by** X
having set(A) **contains-none** $\{v_{1,1}, \ldots, v_{1,j_1}\}$ **and if possible** …
and if possible $\{v_{n,1}, \ldots, v_{n,j_n}\}$.

It is worth noticing that an expression based on one (or several) difference(s) would be complicated to formulate and thus would not be natural at all (especially for a user), while the one chosen above is. Moreover, the query specifies dislikes which are given in a hierarchical manner. So, S_1 contains the values the most highly (indeed totally excluded) and S_n those which are the most weakly unwanted. Here also, associated with the preference relation sustaining the hierarchy, is an ordinal scale L with labels l_i's (such that $l_1 > \ldots > l_n > l_{n+1}$) which will be used to assign levels of satisfaction to the elements of the result of any stratified anti-division.

Example 5. Let us consider the case of a consumer who wants food products (e.g., noodles or vegetal oil) without certain additive substances. In the presence of the relation products(p-name, add-s) describing which additives (add-s) are involved in products, a possible query is:

select top 5 p-name **from** products **group by** p-name
having set(add-s) **contains-none** {AS27, BT12, C3}
and if possible {AS5, D2} **and if possible** {D8}

which tells that the additives AS27, BT12 and C3 are completely forbidden, that the absence of both AS5 and D2 is appreciated and that it is still better if D8 is not in the product. ♦

3.3 Modeling Stratified Operations

We consider a stratified division or anti-division of a relation r whose schema is (A, X) by a relation s defined over attribute B with A and B compatible attributes (in fact, A and B could be compatible sets of attributes as well). The principle for defining these operations is to extend expressions (2) and (5). This point of departure entails: i) dealing with the preferences applying to the divisor and ii) using an ordinal (symbolic) implication. This is why we use an augmented relational framework where each tuple of a relation rel is assigned a (symbolic) level of preference taken from the scale L, denoted by $pref_{rel}(t)$ and any tuple can be written $pref_{rel}(t)/t$. Since the dividend relation is not concerned with explicit preferences, its tuples are assigned the maximal level l_1 while the tuples which are absent are (virtually) assigned the worst level l_{n+1}. For the divisor, the level of preference attached to a tuple is directly stemming from the place of the corresponding element in the hierarchy provided by the user. As to the implication, it can be chosen among fuzzy implications with two requirements: i) to work in a purely ordinal context, and ii) to convey the semantics of importance associated with the layered divisor. It turns out that Kleene-Dienes implication usually defined as:

$$p \Rightarrow_{KD} q = \max(1 - p, q)$$

meets the goal provided that the complement to 1 is changed into order reversal over L. In other words, we will use a symbolic version of the previous implication, denoted by $\Rightarrow_{sKD}$:

$$l_i \Rightarrow_{sKD} l_j = \max(\text{rev}(l_i), l_j)$$

where $\forall l_i \in L = l_1 > \ldots l_{n+1}$, $\text{rev}(l_i) = l_{n+2-i}$. In other words, if a symbol s has the position k on the scale, rev(s), its negation, has the position k when the scale is read from the end.

Example 6. Let L be the scale:

completely important > highly important > fairly important >
not very important > not at all important.

The inverse scale is :

[rev(completely important) = not at all important] <
[rev(highly important) = not very important] <
[rev(fairly important) = fairly important] <
[rev(not very important) = highly important] <
[rev(not at all important) = completely important]. ♦

So equipped, if V denotes the values of the divisor, the stratified division and anti-division are defined as follows:

$$\begin{aligned} pref_{\text{strat-div}(r, s, A, B)}(x) &= \min_{v \in V} pref_s(v) \Rightarrow_{sKD} pref_r(v, x) \\ &= \min_{v \in V} \max(\text{rev}(pref_s(v)), pref_r(v, x)) \end{aligned} \quad (7)$$

$$\begin{aligned} pref_{\text{strat-antidiv}(r, s, A, B)}(x) &= \min_{v \in V} pref_s(v) \Rightarrow_{sKD} \text{rev}(pref_r(v, x)) \\ &= \min_{v \in V} \max(\text{rev}(pref_s(v)), \text{rev}(pref_r(v, x))). \end{aligned} \quad (8)$$

Due to the fact that $\text{pref}_r(v, x)$ takes only the two values l_1 and l_{n+1} depending on the presence or absence of <v, x> in relation r:

i) in expression (7), each term $\max(\text{rev}(\text{pref}_V(v)), \text{pref}_r(v, x))$ equals l_1 if x is associated with v in r ($<v, x> \in r$), $\text{rev}(\text{pref}_V(v))$ otherwise,
ii) in expression (8), each term $\max(\text{rev}(\text{pref}_V(v)), \text{rev}(\text{pref}_r(v, x)))$ equals l_1 if x is not associated with v in r ($<v, x> \notin r$), $\text{rev}(\text{pref}_V(v))$ otherwise.

In other words, if x is associated with all (respectively none) of the values of the entire divisor, the maximal level of preference l_1 is obtained and as soon as an association <v, x> is missing (respectively found), the level of preference of x decreases all the more as v is highly preferred (respectively undesired).

Example 7. Let us consider the following dividend relation r:

$$r = \{<a1, x>, <a2, x>, <a4, x>, <a1, y>, <a3, y>, <a5, z>, <a2, t>\}$$

and the stratified divisor:

$$s = \{a1\} \succ \{a2\} \succ \{a3, a4\}$$

which induces the four-level scale $L = l_1 > l_2 > l_3 > l_4$. These relations rewrite:

r

A	X	pref
a1	x	l_1
a2	x	l_1
a4	x	l_1
a1	y	l_1
a3	y	l_1
a2	z	l_1
a3	z	l_1
a4	z	l_1
a2	t	l_1

s

B	pref
a1	l_1
a2	l_2
a3	l_3
a4	l_3

According to formula (7), the result d-res of the division of r by s is:

$\text{pref}_{\text{d-res}}(x) = \min(l_1, l_1, \text{rev}(l_3), l_1) = l_2$,

$\text{pref}_{\text{d-res}}(y) = \min(l_1, \text{rev}(l_2), l_1, \text{rev}(l_3)) = l_3$,

$\text{pref}_{\text{d-res}}(z) = \min(\text{rev}(l_1), \text{rev}(l_2), \text{rev}(l_3), \text{rev}(l_3)) = l_4$,

$\text{pref}_{\text{d-res}}(t) = \min(\text{rev}(l_1), l_1, \text{rev}(l_3), \text{rev}(l_3)) = l_4$,

which means that x is preferred to y on the one hand and that z and t are rejected on the other hand. Similarly, using formula (8), the following result ad-res of the anti-division of r by s is obtained:

$\text{pref}_{\text{ad-res}}(x) = \min(\text{rev}(l_1), \text{rev}(l_2), l_1, \text{rev}(l_3)) = l_4$,

$\text{pref}_{\text{ad-res}}(y) = \min(\text{rev}(l_1), l_1, \text{rev}(l_3), l_1) = l_4$,

$\text{pref}_{\text{ad-resr}}(z) = \min(l_1, l_1, l_1, l_1) = l_1,$

$\text{pref}_{\text{ad-res}}(t) = \min(l_1, \text{rev}(l_2), l_1, l_1) = l_3,$

which states that z is fully satisfactory and t significantly less, while x and y are quite unsatisfactory. ♦

3.4 *Property of the Result of Stratified Divisions and Anti-divisions*

In order to be qualified a division (respectively anti-division), the extended operator defined above must deliver a result having the characteristic property of a quotient. This means that one must have valid properties similar to 3a-b (respectively 6a-b). In [2], it is shown that the division of fuzzy relations (i.e., where each tuple is assigned a membership degree taken in the unit interval) leads to a result which is a quotient as far as the implication used is either an R-implication, or an S-implication. The key point of the proof lies in the fact that these implications ($\Rightarrow_f$) may be written in a common format, namely :

$$p \Rightarrow_f q = \sup \{y \in [0, 1] \mid \text{cnj}(p, y) \leq q\}$$

where cnj is an appropriate conjunction operator (see [2, 7] for more details). In the specific case considered here, the ordinal version of Kleene-Dienes implication (which belongs to the family of S-implications) writes:

$$l_i \Rightarrow_{sKD} l_j = \max(\text{rev}(l_i), l_j) = \sup \{y \in [l_1, l_{n+1}] \mid \text{cnj}(l_i, y) \leq l_j\} \quad (9)$$

with $\text{cnj}(a, b) = l_{n+1}$ if $a \leq \text{rev}(b)$, b otherwise. (10)

So, if we denote by d-res (repectively ad-res) the result of a stratified division (respectively anti-division), due to the very nature of expression (9), the following expressions hold:

$s \times \text{d-res} \subseteq r$ (11a) $\forall \text{d-res}' \supset \text{d-res},\ s \times \text{d-res}' \not\subseteq r$ (11b)

$s \times \text{ad-res} \subseteq \text{cp}(r)$ (12a) $\forall \text{ad-res}' \supset \text{ad-res},\ s \times \text{ad-res}' \not\subseteq \text{cp}(r)$ (12b)

where the Cartesian product (×), inclusion and complement are respectively defined as:

$r \times s = \{p3/uv \mid p1/u \in r \wedge p2/\ v \in s> \wedge\ p3 = \text{cnj}(p1, p2)\},$

$r \subseteq s \Leftrightarrow \forall\ p1/u \in r, \exists\ p2/u \in s$ such that $p1 \leq p2,$

$\text{cp}(r) = \{\text{rev}(p)/u \mid p/u \in r\}$

which means that d-res is a quotient and that ad-res is an anti-quotient.

Example 8. Let us come back to the relations of example 7. According to (11a), we must have:

s

B	pref
a1	l_1
a2	l_2
a3	l_3
a4	l_3

×

d-res

X	pref
x	l_2
y	l_3

⊆

r

A	X	pref
a1	x	l_1
a2	x	l_1
a4	x	l_1
a1	y	l_1
a3	y	l_1
a2	z	l_1
a3	z	l_1
a4	z	l_1
a2	t	l_1

on the one hand, and with respect to (11b), if any grade is increased in d-res (yielding d-res'), the Cartesian product of s and d-res' is not included in r. We will illustrate what happens for x and t (which may be considered to be in d-res with the level of preference l_4) and it would be easy to observe that the same conclusions can be drawn for y and z.

In the Cartesian product, we have the tuples:

$cnj(l_1, l_2)/<a1, x> = l_2/<a1, x>$,
$cnj(l_2, l_2)/<a2, x> = l_2/<a2, x>$,
$cnj(l_3, l_2)/<a3, x> = l_4/<a3, x>$,
$cnj(l_3, l_2)/<a4, x> = l_4/<a4, x>$,
$cnj(l_1, l_4)/<a1, t> = l_4/<a1, t>$,
$cnj(l_2, l_4)/<a2, t> = l_4/<a2, t>$,
$cnj(l_3, l_4)/<a3, t> = l_4/<a3, t>$,
$cnj(l_3, l_4)/<a4, t> = l_4/<a4, t>$,

and the inclusion in r holds. If we suppose that the level of preference of x in d-res is increased (from l_2 to l_1), the partial Cartesian product of s and l_1/x yields:

$cnj(l_1, l_1)/<a1, x> = l_1/<a1, x>$,
$cnj(l_2, l_1)/<a2, x> = l_1/<a2, x>$,
$cnj(l_3, l_1)/<a3, x> = l_1/<a3, x>$,
$cnj(l_3, l_1)/<a4, x> = l_1/<a4, x>$,

for which the inclusion in r does not hold (presence of the tuple $l_1/<a3, x>$ which does not belong to r). Similarly, let us increase the level of preference of t in d-res (from l_4 to l_3), the partial Cartesian product of s and l_3/t is:

$cnj(l_1, l_3)/<a1, t> = l_3/ <a1, t>$,
$cnj(l_2, l_3)/<a2, t> = l_4/<a2, t>$,
$cnj(l_3, l_3)/<a3, t> = l_4/<a3, t>$,
$cnj(l_3, l_3)/<a4, t> = l_4/<a4, t>$,

and the tuple l_3, $<a1, t>$ violates the inclusion in r. Due to the increasing monotonicity of cnj with respect to its second argument, any other increase of the level of preference of t in d-res would also lead to the non inclusion of the Cartesian product in r.

We now consider the anti-division of r by s and, for illustration purpose, only the elements l_4/y and l_1/z of its result ad-res. In order to check formula (12a), the Cartesian product of s and these two tuples has to be performed, which results in:

$cnj(l_1, l_4)/<a1, y> = l_4/<a1, y>$,
$cnj(l_2, l_4)/<a2, y> = l_4/<a2, y>$,
$cnj(l_3, l_4)/<a3, y> = l_4/<a3, y>$,
$cnj(l_3, l_4)/<a4, y> = l_4/<a4, y>$,
$cnj(l_1, l_1)/<a1, z> = l_1/<a1, z>$,
$cnj(l_2, l_1)/<a2, z> = l_1/<a2, z>$,
$cnj(l_3, l_1)/<a3, z> = l_1/<a3, z>$,
$cnj(l_3, l_1)/<a4, z> = l_1/<a4, z>$,

and the inclusion in the complement of r holds. As to the satisfaction of (12b), clearly the level of preference of z (l_1) is maximal and if that of y is increased from l_4 to l_3, we have the Cartesian product:

$cnj(l_1, l_3)/<a1, y> = l_3/<a1, y>$,
$cnj(l_2, l_3)/<a2, y> = l_4/<a2, y>$,
$cnj(l_3, l_3)/<a3, y> = l_4/<a3, y>$,
$cnj(l_3, l_3)/<a4, y> = l_4/<a4, y>$,

and the tuple $l_3/<a1, y>$ violates the desired inclusion ($l_3 > rev(l_1) = l_4$). ♦

4 Stratified Queries Mixing Division and Anti-division Features

4.1 A Basis for Safe Mixed Queries

The starting point of this section is the analogy between division queries and the search for documents indexed by a certain set of keywords, since these two activities are concerned with the association of an element (respectively a document) with a set of values (respectively keywords). On this line, it seems convenient to extend/enhance the basis of document retrieval with a set of undesired keywords, which has a direct counterpart in terms of anti-division. Last, if we introduce the notion of levels of importance of the keywords in both the positive and negative parts, we end up with a query involving a stratified division (corresponding to the desired keywords/positive part) and a stratified anti-division (corresponding to the unwanted keywords/negative part). Consequently, in the following, we consider queries where the association and non association conditions relate to a same attribute, even it would make sense to envisage more general situations.

A query is made of two parts: i) the positive part which gathers the values which are desired (at different levels of importance) and ii) the negative part which collects the unwanted values, still with different importances. In fact, such queries call on two types of bipolarity: i) one tied to the fact that some conditions (the association with all (respectively none) of the values of the first set) are mandatory whereas others (the association (respectively non association) with the values of the next sets) are only desirable, and ii) another related to the fact that the association with some values is expected (those of the positive part), while one would like the non association with other values (those of the negative part). Clearly, these two types of bipolarity impact the semantics of a query in two quite different ways. The first one entails handling the associations (respectively non associations) with the values of the first stratum as constraints (whose satisfaction or not causes acceptance or rejection) and the associations (respectively non associations) with the values of the other layers as wishes (whose satisfaction or not influences the discrimination between selected elements). The second aspect leads to distinguish between values which are desired and values which are unwanted, then to look for the association with the former ones and for the non association with the latter ones.

A first approach to mixed stratified queries is to consider them as made of two components according to the following pattern:

select top k X **from** r [**where** condition] **group by** X
having set(A) **contains** $\{v_{1,1}, \ldots, v_{1,j_1}\}$ **and if possible** …
and if possible $\{v_{n,1}, \ldots, v_{n,j_n}\}$ **and**
contains-none $\{w_{1,1}, \ldots, w_{1,k_1}\}$ **and if possible** …
and if possible $\{w_{p,1}, \ldots, w_{p,k_p}\}$.

This means that the query refers to two scales:

$L1 = l_1 > \ldots > l_n > l_{n+1}$ for the positive part

and:

$L2 = l'_1 > \ldots > l'_p > l'_{p+1}$ for the negative part

and the overall satisfaction of a given x would require to combine two symbols (one from each scale), which raises a serious problem.

To avoid this difficulty, we suggest to build mixed queries in such a way that a single scale comes into play. Each level of the scale used in a query will be assigned a set of desired values (contributing the positive part) and a set of unwanted values (subset of the negative part), one of them being possibly empty. A mixed stratified query will be expressed according to the following model:

select top k X **from** r [**where** condition] **group by** X
having set(A) **contains** [**pos**: $\{v_{1,1}, \ldots, v_{1,j_1}\}$, **neg**: $\{w_{1,1}, \ldots, w_{1,k_1}\}$]
and if possible …
and if possible [**pos**: $\{v_{n,1}, \ldots, v_{n,j_n}\}$, **neg**: $\{w_{n,1}, \ldots, w_{n,k_n}\}$]

where "pos (respectively neg): S", at a given level of importance, stands for a set of desired (respectively unwanted) values, which x must be (repectively not be) associated with. In addition, note that it is possible to have "pos : {}", "neg : {}" (but not both) at each layer.

4.2 Syntax, Semantics and Modeling of Mixed Queries

The above type of query is interpreted in a straightforward manner as follows. To be somewhat satisfactory, a element x: i) must be associated with all the values $\{v_{1,1}, \ldots, v_{1,j_1}\}$ and none of the values of $\{w_{1,1}, \ldots, w_{1,k_1}\}$, and ii) it receives a level of satisfaction (pref) all the larger as it satisfies the association with all the values $\{v_{2,1}, \ldots, v_{2,k_2}\}, \ldots, \{v_{j,1}, \ldots, v_{j,k_j}\}$ and none of the values of $\{w_{2,1}, \ldots, w_{2,p_2}\}, \ldots, \{w_{j,1}, \ldots, w_{j,p_j}\}$ with j taking a high value (n for the maximal level pref = l_1). In other words, an element x is preferred to another y if x is connected with $\{v_{1,1}, \ldots, v_{1,k_1}\}, \ldots, \{v_{i,1}, \ldots, v_{i,k_i}\}$ and none of $\{w_{1,1}, \ldots, w_{1,p_1}\}, \ldots, \{w_{i,1}, \ldots, w_{i,p_i}\}$, while y is associated with $\{v_{1,1}, \ldots, v_{1,k_1}\}, \ldots, \{v_{j,1}, \ldots, v_{j,k_j}\}$ and none of $\{w_{1,1}, \ldots, w_{1,p_2}\}, \ldots, \{w_{j,1}, \ldots, w_{j,p_j}\}$ and i > j.

Let us denote by s = $\{V_1, \ldots, V_n\}$ the different layers of the divisor where each V_i is made of a positive part P_i and a negative part N_i. Alternatively, s writes as s = (P, N), its positive and negative parts. The mixed stratified division is defined as:

$$\begin{aligned}
&\text{pref}_{\text{mix-strat-div}(r,\, s,\, A,\, B)}(x) = \\
&\quad \min_{i \in [1, n]} \min(\min_{v \in P_i} \text{pref}_s(v) \Rightarrow_{sKD} \text{pref}_r(v, x)), \\
&\quad\quad \min_{w \in N_i} \text{pref}_s(w) \Rightarrow_{sKD} \text{rev}(\text{pref}_r(w, x))) \\
&= \min_{i \in [1, n]} \min(\min_{v \in P_i} \max(\text{rev}(l_i), \text{pref}_r(v, x)), \\
&\quad\quad \min_{w \in N_i} \max(\text{rev}(l_i), \text{rev}(\text{pref}_r(w, x)))) \\
&= \min(\min_{v \in P} \max(\text{rev}(\text{pref}_s(v)), \text{pref}_r(v, x)), \\
&\quad\quad \min_{w \in N} \max(\text{rev}(\text{pref}_s(w)), \text{rev}(\text{pref}_r(w, x)))). \qquad (13)
\end{aligned}$$

By construction, the result delivered by the above operation is a quotient in the sense that it is a maximal relation. More precisely, it is the largest (ordinal) relation whose Cartesian product (using the conjunction given in expression (10)) with the positive and negative parts of the divisor is included in the dividend. So, if m-res denotes the result delivers by expression (13), the following characterization formulas hold:

$$\left.\begin{array}{l} P \times \text{m-res} \subseteq r \\ \quad\text{and} \\ N \times \text{m-res} \subseteq \text{cp}(r) \end{array}\right\} \qquad (14a)$$

$$\forall \text{m-res}' \supset \text{m-res}, \quad \left.\begin{array}{l} P \times \text{m-res}' \not\subseteq r \\ \quad\text{or} \\ N \times \text{m-res}' \not\subseteq \text{cp}(r) \end{array}\right\} \qquad (14b)$$

4.3 A Complete Example

Let us consider a relation Prod(product, component) where a tuple <p, c> expresses that c is one of the components of product p and the mixed division query:

select top 5 product **from** Prod **group by** product
having set(product) **contains** [**pos**: $\{c_1\}$, **neg**: $\{c_5, c_6\}$]
and if possible [**pos**: $\{c_2\}$, **neg**: $\{\}$]
and if possible [**pos**: $\{c_3, c_4\}$, **neg**: $\{c_7\}$]

expressing that the double stratification:

$$P: \{c_1\}\ (l_1) > \{c_2\}\ (l_2) > \{c_3, c_4\}\ (l_3)$$
$$N: \{c_5, c_6\}\ (l_1) > \varnothing > \{c_7\}\ (l_3).$$

The user wants product c_1, if possible c_2 and if possible c_3 and c_4, and he/she dislikes c_5 and c_6 (respectively c_7) as much as he/she desires c_1 (respectively c_3 and c_4). Notice that there is no counterpart for c_2 (in other words c_3 and c_4 are forbidden and c_7 is only weakly undesired). If the dividend relation is:

$$r = \{<c_1, x>, <c_2, x>, <c_4, x>, <c_1, y>, <c_2, y>, <c_3, y>, <c_4, y>, <c_7, y>, <c_2, z>, <c_3, z>, <c_4, z>, <c_1, t>, <c_2, t>, <c_5, t>, <c_1, u>\}.$$

According to formula (13), the levels of satisfaction assigned to x, y, z and t are:

$$\begin{aligned}
\text{pref}(x) = \min(&\min(l_1 \Rightarrow_{sKD} \text{pref}_r(c_1, x), l_2 \Rightarrow_{sKD} \text{pref}_r(c_2, x), \\
&l_3 \Rightarrow_{sKD} \text{pref}_r(c_3, x), l_3 \Rightarrow_{sKD} \text{pref}_r(c_4, x)), \\
&\min(l_1 \Rightarrow_{sKD} \text{rev}(\text{pref}_r(c_5, x)), l_1 \Rightarrow_{sKD} \text{rev}(\text{pref}_r(c_6, x)), \\
&l_3 \Rightarrow_{sKD} \text{rev}(\text{pref}_r(c_7, x))) \\
= \min(&l_1, l_1, l_2, l_1, l_1, l_1, l_1) = l_2
\end{aligned}$$

$$\begin{aligned}
\text{pref}(y) = \min(&\min(l_1 \Rightarrow_{sKD} \text{pref}_r(c_1, y), l_2 \Rightarrow_{sKD} \text{pref}_r(c_2, y), \\
&l_3 \Rightarrow_{sKD} \text{pref}_r(c_3, y), l_3 \Rightarrow_{sKD} \text{pref}_r(c_4, y)), \\
&\min(l_1 \Rightarrow_{sKD} \text{rev}(\text{pref}_r(c_5, y)), l_1 \Rightarrow_{sKD} \text{rev}(\text{pref}_r(c_6, y)), \\
&l_3 \Rightarrow_{sKD} \text{rev}(\text{pref}_r(c_7, y))) \\
= \min(&l_1, l_1, l_1, l_1, l_1, l_1, l_2) = l_2
\end{aligned}$$

$$\begin{aligned}
\text{pref}(z) = \min(&\min(l_1 \Rightarrow_{sKD} \text{pref}_r(c_1, z), l_2 \Rightarrow_{sKD} \text{pref}_r(c_2, z), \\
&l_3 \Rightarrow_{sKD} \text{pref}_r(c_3, z), l_3 \Rightarrow_{sKD} \text{pref}_r(c_4, z)), \\
&\min(l_1 \Rightarrow_{sKD} \text{rev}(\text{pref}_r(c_5, z)), l_1 \Rightarrow_{sKD} \text{rev}(\text{pref}_r(c_6, z)), \\
&l_3 \Rightarrow_{sKD} \text{rev}(\text{pref}_r(c_7, z))) \\
= \min(&l_4, l_1, l_1, l_1, l_1, l_1, l_1) = l_4
\end{aligned}$$

$$\begin{aligned}
\text{pref}(t) = \min(&\min(l_1 \Rightarrow_{sKD} \text{pref}_r(c_1, t), l_2 \Rightarrow_{sKD} \text{pref}_r(c_2, t), \\
&l_3 \Rightarrow_{sKD} \text{pref}_r(c_3, t), l_3 \Rightarrow_{sKD} \text{pref}_r(c_4, t)), \\
&\min(l_1 \Rightarrow_{sKD} \text{rev}(\text{pref}_r(c_5, t)), l_1 \Rightarrow_{sKD} \text{rev}(\text{pref}_r(c_6, t)), \\
&l_3 \Rightarrow_{sKD} \text{rev}(\text{pref}_r(c_7, t))) \\
= \min(&l_1, l_1, l_2, l_2, l_4, l_1, l_1) = l_4
\end{aligned}$$

$$\begin{aligned}
\text{pref}(u) = \min(&\min(l_1 \Rightarrow_{sKD} \text{pref}_r(c_1, u), l_2 \Rightarrow_{sKD} \text{pref}_r(c_2, u), \\
&l_3 \Rightarrow_{sKD} \text{pref}_r(c_3, u), l_3 \Rightarrow_{sKD} \text{pref}_r(c_4, u)), \\
&\min(l_1 \Rightarrow_{sKD} \text{rev}(\text{pref}_r(c_5, u)), l_1 \Rightarrow_{sKD} \text{rev}(\text{pref}_r(c_6, u)), \\
&l_3 \Rightarrow_{sKD} \text{rev}(\text{pref}_r(c_7, u))) \\
= \min(&l_1, l_3, l_2, l_2, l_1, l_1, l_1) = l_3.
\end{aligned}$$

Finally, one has the resulting relation: $\{l_2/x, l_2/y, l_3/u\}$.

It turns out that x and y are equally ranked since the absence of $<c_3, x>$ has the same impact for x as the presence of $<c_7, y>$ for y (the level of desire for c_3 equals the level of dislike for $c_7 - l_3$). Similarly, the absence of c_1 (mandatory) for z has the same effect (rejection) as the presence of c_5 (forbidden) for t.

Using the non commutative conjunction defined in formula (10) (and discarding the tuples whose level is l_4), the Cartesian product of the positive part of the divisor and $\{<l_2, x>, <l_2, y>, <l_3, u>\}$ is:

$$\{l_1/<c_1, x>, l_1/<c_2, x>, l_2/<c_1, y>, l_2/<c_2, y>, l_3/<c_1, u>\}$$

which is included in the dividend r. Similarly, the Cartesian product of the negative part of the divisor and the previous result yields:

$$\{l_2/<c_5, x>, l_2/<c_6, x>, l_2/<c_5, y>, l_2/<c_6, y>, l_3/<c_5, u>, l_3/<c_6, u>\}$$

which is included in the complement of the dividend (i.e., none of these tuples appears in the dividend). We observe that formula (14a) holds.

It is easy to check that if the level of preference of any element (x, y, z, t or u) is upgraded, the property conveyed by formula (14b) is valid. For instance, if we consider l_1/x instead of l_2/x, the Cartesian product (with P) becomes:

$$\{l_1/<c_1, x>, l_1/<c_2, x>, l_1/<c_3, x>, l_1/<c_4, x>, l_2/<c_1, y>, l_2/<c_2, y>, l_3/<c_1, u>\}$$

and the presence of the third tuple shows the non-inclusion in the dividend. Similarly, if the tuple l_3/z is introduced, the Cartesian product (with P) becomes:

$$\{l_1/<c_1, x>, l_1/<c_2, x>, l_2/<c_1, y>, l_2/<c_2, y>, l_3/<c_1, u>, l_3/<c_1, z>\}$$

and the last tuple proves that the inclusion in r does not hold (then that property (14b) is valid). Last, if l_2/y is replaced by l_1/y, the Cartesian product of N and the modified result is:

$$\{l_2/<c_5, x>, l_2/<c_6, x>, l_1/<c_5, y>, l_1/<c_6, y>, l_1/<c_7, y>, l_3/<c_5, u>, l_3/<c_6, u>\}$$

and the presence of the tuple $l_1/<c_7, y>$ makes the inclusion in the dividend fail, which, once again, shows the validity of property (14b). ♦

5 Implementation Issues

Now, we tackle processing strategies issues for division and anti-division queries. The objective is to suggest several algorithms which are suited to a reasonably efficient evaluation of such queries (subsections 5.1 and 5.2) and to assess the extra cost with respect to queries involving no preferences (subsection 5.3).

5.1 Processing of Division Queries

Three algorithms implementing formula 7 are successively described. The first algorithm is based on a sequential scan of the dividend (SSD). The idea is to access the tuples from the dividend relation (r) "in gusts", i.e., by series of tuples which share the same X-attribute value (in the spirit of what is performed by a "group by" clause). Moreover, inside a cluster the tuples (x, a) are ordered increasingly on A. This is performed by the query:

select * **from** r **order by** X, A.

Thanks to a table which gives, for each value (val-A) of the divisor, the layer to which it belongs (str-A), one can update the number of values from each layer which are associated with the current element x, while scanning the result of the query above. At the end of a group of tuples, one checks the layers in decreasing order of their importance. The process stops as soon as the current element x is not associated with all of the values from a layer V_i. Three cases can appear: i) x is associated with all of the values from all the layers of the divisor and it gets the preference level l_1, ii) the stop occurs while checking layer V_i whose importance is not maximal (i > 1) and x gets the preference level $rev(l_i) = l_{n+2-i}$, iii) the stop occurs while checking layer V_1 and x gets the level l_{n+1} meaning that it is rejected.

In the second algorithm, data accesses are guided by the divisor (AGD). Thus, instead of scanning the dividend exhaustively and then checking the layers satisfied by a given x by means of the aforementioned table, one first retrieves the X-values from the dividend, and for each such x, the associations with the different layers are checked by means of an SQL query involving the aggregate count. Again, a layer is checked only if the layers of higher importance had all of their values associated with x. The first step is to retrieve the distinct values of attribute X present in r by means of the query:

select distinct X **from** r.

Then, for each value x returned, one counts the A-values from V_1 which are associated with x (whose current value is denoted by :x below) in r by means of the query:

select count(*) **from** r **where** X = :x **and** A **in** (**select** A **from** V_1).

If the value returned equals the cardinality of V_1, one checks layer V_2 by means of a similar query, and so on. The loop stops as soon as a missing association with the current layer is detected. The preference level assigned to x is computed according to the same principle as in the previous algorithm.

The last strategy relies on a series of regular division queries (SRD). It consists of two steps: i) to process as many regular division queries as there are layers in the divisor, and ii) to merge the different results and compute the final preference degrees. The algorithm has the following general shape:

step 1: for each layer V_i of the divisor, one processes a division query which retrieves the x's which are associated in r with all of the values from V_i. The layers are examined in decreasing order of their importance and an element x is checked only if it belongs to the result of the query related to the previous layer.

step 2: the results $T_1, \ldots, T_n$ of the previous division queries are merged by taking them in decreasing order of the corresponding layers. An element x which belongs to T_i (the result of layer V_i) but not to T_{i+1} gets the preference level l_{n-i+1} (assuming that there exists a table T_{n+1} which is empty). We have used an algorithm where the query (in step 2) rests on an outer join.

5.2 Processing of Anti-division Queries

Each of the previous methods can be adapted so as to apply to anti-division queries. In the SSD algorithm, after running the query:

select * **from** r **order by** X, A,

using the table connecting each value of the divisor with its layer, it is possible to identify the occurrence(s) of unwanted values. At the end of a cluster of tuples, the level of preference assigned to the current element x is determined by checking the layers in decreasing order of their importance. Here also, the process can stop as soon as the current element x is associated with one of the values from a layer V_i (x receives the level l_{n+1} if $i = 1$, l_{n+2-i} if $i \in [2, n]$) and if no undesired association is detected, x is assigned the level l_1.

Similarly, the algorithm AGD is transformed as follows. As originally, the distinct values of attribute X present in r are retrieved by means of the query:

select distinct X **from** r.

Then, for each value x, the number of A-values from V_1 (the totally excluded values specified in the divisor) which are associated with x in r, is computed by means of the query:

select count(*) **from** r **where** X = :x **and** A **in** (**select** A **from** V_1).

If the value returned is zero, one checks layer V_2 by means of a similar query, and so on. The loop stops as soon as an unwanted association with the current layer is detected. The preference level assigned to x is computed according to the same principle as in the previous algorithm.

The strategy SRD now means "a series of regular differences". The first step rests on queries of type:

(**select** X **from** r) **differ** (**select** X **from** r **where** A **in** (**select** B **from** V_i))

for each set V_i corresponding to a layer of the divisor. The second step takes all the successive pairs of results produced previously in order to assign the preference level l_{n-i+1} to an element x which belongs to the result of layer V_i but not to that of layer V_{i+1}.

5.3 Experiments

As mentioned previously, the objectives of the experimentation are mainly to assess the additional processing cost related to the handling of preferences and to compare the performances of the algorithms presented above. The experimentation was performed with the DBMS Oracle™ Enterprise Edition Release 8.0.4.0.0 running on an Alpha server 4000 bi-processor with 1.5 Gb memory.

A generic stratified division (respectively anti-division) query has been run on dividend relations of 300, 3000 and 30000 tuples, and a divisor including five layers made of respectively 3, 2, 1, 2 and 2 values. The query taken as a reference is the analogous division (respectively anti-division) query without preferences, where the divisor is made of the sole first layer of the divisor (which corresponds to a "hard constraint" as mentioned before). So doing, we can assess the extra cost related only to the "preference part" of the query, i.e., to the presence of the non-mandatory layers. The reference division query has been evaluated using three methods: i) sequential scan of the dividend (i.e., algorithm SSD without preferences, denoted by REF1), ii) access guided by the divisor (i.e., algorithm AGD without preferences, denoted by REF2), iii) algorithm REF3 based on a query involving a "group by" clause and a counting, as in the first step of algorithm SRD. The reference anti-division query has been evaluated using these same methods. However, it is worth notocing that REF1 shows the same performances for both the division and anti-division since the only difference lies in the final comparison of the cardinality of the current subset (with that of the layer for the division and with 0 for the anti-division. Moreover:

- we used synthetic data generated in such a way that the selectivity of each value b from the divisor relatively to any x from the dividend is equal to 75% in the case of a division query (for a given value b from the divisor and a given x from the dividend, tuple (x, b) has three chances out of four to be present in the dividend), and it is equal to 25% in the case of an anti-division query,
- each algorithm was run 8 times so as to avoid any bias induced by the load of the machine,
- the time unit equals 1/60 second.

The results obtained for the division are reported in the table hereafter:

Size of the dividend	300	3000	30000
REF1	15.8	144.6	1451
REF2	49.7	570	15536
REF3	11.4	40.5	361.9
SSD	99	1011	10451
AGD	84	1035	29927
SRD	89	332	2923
Number of answers	15	172	1693

One can notice that:

- among the reference methods for non-stratified operations, the most efficient is by far REF3. This is due to the fact that is is based on a single query involving a "group by" clause, which is very efficiently optimized by the system,

- the processing time of the algorithms based on a sequential scan of the dividend (i.e., SSD and REF1) vary linearly w.r.t. the size of the dividend, contrary to those from the second family (REF2 and AGD); as to algorithm SRD (implemented with an outer join), its complexity shows some linearity as soon as the size of the dividend is above a certain threshold (which means that there is a fixed cost attached to it, which depends on the number of layers of the divisor),
- algorithm SSD becomes better than AGD as soon as the size of the dividend is over 1000 tuples. But the best algorithm is SRD (implemented with an outer join), which outperforms all the others as soon as the dividend contains more than 300 tuples. It is worth noticing that the ratio between SRD and REF3 is almost constant (around 8), which is due to the fact that SRD performs one query of type REF3 per layer (here 5), plus the combination of the intermediate results.

To sum up, it appears that algorithm SRD based on an outer join is the most efficient, except for very small sizes of the dividend where AGD is slightly better. However, the extra cost of SRD with respect to the most efficient reference algorithm, namely REF3, is still important (the multiplicative factor is around 8).

The results obtained for the anti-division are reported in the next table:

Size of the dividend	300	3000	30000
REF1	15.8	144.6	1451
REF2	41.4	400.7	4055
REF3	13.2	81.4	760.2
SSD	108.6	960.5	10418
AGD	54.2	645.2	6315
SRD	106	375.1	4353
Number of answers	37	427	4365

These results show that:

- among the reference methods for non-stratified anti-divisions, REF3 is much more efficient than REF2; the fact that it outperforms REF2 by such a large margin means that the DBMS is not efficient at optimizing nested queries,
- the performances of REF2, AGD and SSD vary linearly with respect to the size of the dividend. As to REF3 and SRD, their complexity is less than linear.

It turns out that the best algorithm for stratified anti-divisions is SRD, which is significantly better than AGD, itself much more efficient than SSD. However, the extra cost of SRD with respect to the most efficient reference algorithm, namely REF3, is still rather important (multiplicative factor between 4.6 and 8).

What all these measures show was somewhat predictable: the best way to process a division or anti-division query (stratified or not) is to express it by means of a single query that can be efficiently handled by the optimizor of the system, and not by external programs which induce a more or less important overhead. For instance, in the case of the anti-division, the extra cost attached to SRD with respect to REF3 is explainable by the fact that SRD processes five regular anti-division queries (one for each layer) instead of one for REF3, and then has to merge the results of these queries. Consequently, if the stratified division or anti-division functionality were to be integrated into a commercial DBMS, it is clear that it would have to be handled by the optimizor at an internal level, and processed as one query, according to the format given in Subsection 3.2 and in such a way that the evaluation of a given x is done in one step.

6 Conclusion

In this article, we dealt with division and anti-division queries involving user preferences expressed in an ordinal way. The principle consists in using a divisor made of a hierarchy of layers. The first layer corresponds to a set of mandatory values, whereas the other layers are used to discriminate among the elements in the result. So doing, the result is no longer a flat set but a list of items provided with a level of satisfaction. It has been shown that the stratified division (respectively anti-division) delivers a result that can be characterized as a quotient (respectively an anti-quotient).

Besides, some experimental measures have been carried out in order to assess the feasibility of such extended division or anti-division queries. Even though these measures still need to be completed, they show that the additional cost induced by the stratified nature of the divisor is quite high (multiplicative factor from 5 to 8 with respect to the relative classical operation) but that the overall processing time is still acceptable for medium-sized dividend relations. To reach better performances, it would be of course necessary to integrate the new operator into the processing engine of the system, so as to benefit from a real internal optimization, instead of processing stratified division queries externally, as we did here.

This work opens several perspectives, among which : i) the enrichment of division or anti-division queries whose semantics could be disjunctive with respect to the role of the layers or even based on the lexicographic order as mentioned in the end of subsection 3.1, ii) making complementary experiments in order to take into account larger sizes for both the dividend and the divisor (in particular in the case where the divisor is not specified extensionally by the user, but results from subqueries), iii) the investigation of strategies suited for processing mixed queries in the sense of section 4, along with the corresponding experiments, and iv) checking whether the results obtained with Oracle are confirmed when another DBMS (e.g. PostgresQL or MySQL) is used.

References

1. Börzsönyi, S., Kossmann, D., Stocker, K.: The Skyline operator. In: Proc. of the 17th International Conference on Data Engineering, pp. 421–430 (2001)
2. Bosc, P., Pivert, O., Rocacher, D.: About quotient and division of crisp and fuzzy relations. Journal of Intelligent Information Systems 29, 185–210 (2007)

3. Bosc, P., Pivert, O.: On a parameterized antidivision operator for database flexible querying. In: Bhowmick, S.S., Küng, J., Wagner, R. (eds.) DEXA 2008. LNCS, vol. 5181, pp. 652–659. Springer, Heidelberg (2008)
4. Bouchon-Meunier, B., Dubois, D., Godo, L., Prade, H.: Fuzzy sets and possibility theory in approximate and plausible reasoning. In: Bezdek, J., Dubois, D., Prade, H. (eds.) Fuzzy Sets in Approximate Reasoning and Information Systems, pp. 15–190. Kluwer Academic Publishers, Dordrecht (1999)
5. Bruno, N., Chaudhuri, S., Gravano, L.: Top-k selection queries over relational databases: mapping strategies and performance evaluation. ACM Transactions on Database Systems 27, 153–187 (2002)
6. Chomicki, J.: Preference formulas in relational queries. ACM Transactions on Database Systems 28, 427–466 (2003)
7. Dubois, D., Prade, H.: A theorem on implication functions defined from triangular norms. Stochastica 8, 267–279 (1984); Also in: Dubois, D., Prade, H., Yager, R.R. (eds.) Readings in Fuzzy sets for Intelligent Systems, pp. 105–112. Morgan & Kaufmann, San Francisco (1993)
8. Dubois, D., Prade, H.: Using fuzzy sets in flexible querying: why and how. In: Proc. of the Workshop on Flexible Query-Answering Systems, pp. 89–103 (1996)
9. Dubois, D., Prade, H.: Handling Bipolar Queries in Fuzzy Information Processing. In: Galindo, J. (ed.) Handbook of Research on Fuzzy Information Processing in Databases. Information Science Reference, Hershey (2008)
10. Dubois, D., Prade, H.: An introduction to bipolar representations of information and preference. International Journal of Intelligent Systems 23, 866–877 (2008)
11. Hadjali, A., Kaci, S., Prade, H.: Database preferences queries – A possibilistic logic approach with symbolic priorities. In: Hartmann, S., Kern-Isberner, G. (eds.) FoIKS 2008. LNCS, vol. 4932, pp. 291–310. Springer, Heidelberg (2008)
12. Kießling, W., Köstler, G.: Preference SQL – Design, implementation, experiences. In: Proc. 28th Conference on Very Large Data Bases, pp. 990–1001 (2002)
13. Lacroix, M., Lavency, P.: Preferences: putting more knowledge into queries. In: Proc. 13th Conference on Very Large Data Bases, pp. 217–225 (1987)

Integration of Fuzzy ERD Modeling to the Management of Global Contextual Data

Gregory Vert and S.S. Iyengar

Abstract. This chapter introduces the idiosyncrasies of managing the new paradigm of global contextual data, sets of context data and super sets of context data. It introduces some of the basic idea's behind contexts and then develops a model for management of aggregated sets of contextual data and proposes methods for dealing with the selection and retrieval of context data that is inherently ambiguous about what to retrieve for a given query. Because contexts are characterized by four dimensions, those of time, space, impact and similarity they are inherently complicated to manage.

This work builds on previous work and extends that work to incorporate contexts. The original model for spatial-temporal management is presented and then analyzed to determine much coverage it can provide to the new context paradigm.

Introduction to the Idea of Context

The concept of context has existed in computer science for many years especially in the area of artificial intelligence. The goal of research in this area has been to link the environment a machine exists in to how the machine may process information. An example typically given is that a cell phone will sense that its owner is in a meeting and send incoming calls to voicemail as a result. Application of this idea has been applied to robotics and to business process management [1].

Some preliminary work has been done in the mid 90's. Schilit was one of the first researchers to coin the term context-awareness [2,3]. Dey extended the notion of a context with that of the idea that information could be used to characterize a situation and thus could be responded to [4]. In the recent past more powerful models of contextual processing have been developed in which users are more involved [5]. Most current and previous research has still largely been focused on development of models for sensing devices [6] and not contexts for information processing.

Gregory Vert and S.S. Iyengar
Center For Secure Cyber Security
Louisiana State University
Baton Rouge, LA 70803
e-mail: gvert12@csc.lsu.edu

J. Kacprzyk et al. (Eds.): Uncer. Appro. for Spatial Data Model. and Process., SCI 271, pp. 155–173.
springerlink.com

Little work has been done on the application of contexts to that of how information is processed. The model that we have developed is that of creating metadata describing information events and thus giving them a context. This context then can be used to control the processing and dissemination of such information in a hyper distributed global fashion. The next section will provide a very general overview of the newly developed model and how contexts are defined. The following section will give an overview of the fuzzy ERD model that previously developed could be used for management of contextual information. Finally, the model is evaluated to determine what level of coverage it may provide as it is for management of global contexts data.

Global Contextual Processing

To understand the issues connected with security models for contexts we introduce some details about the newly developing model for contextual processing.

Contextual processing is based on the idea that information can be collected about natural or abstract events and that meta information about the event can then be used to control how the information is processed and disseminated on a global scale. In its simplest form, a context is composed of a feature vector

$$F_n<a_1,..a_n>$$

where the attributes of the vector can be of any data type describing the event. This means that the vector can be composed of images, audio, alpha-numeric etc. Feature vectors can be aggregated via similarity analysis methods into super contexts. The methods that might be applied for similarity reasoning can be statistical, probabilistic (e.g. Baysian), possibilistic (e.g fuzzy sets) or machine learning and data mining based (e.g. decision trees). Aggregation into super sets is done to mitigate collection of missing or imperfect information and to minimize computational overhead when processing contexts.

definition: A context is a collection of attributes aggregated into a feature vector describing a natural or abstract event.

A super context is described as a triple denoted by:

$$S_n = (C_n, R_n, S_n)$$

where C is the context data of multiple feature vectors, R is the meta-data processing rules derived from the event and contexts data and S is controls security processing. S is defined to be a feature vector in this model that holds information about security levels elements or including overall security level requirements.

definition: A super context is a collection of contexts with a feature vector describing the processing of the super context and a security vector that contains security level and other types of security information.

Data Management of Contexts

Having examined contexts, what they contain and how they can be analyzed, it becomes clear that the data management issues of contexts are not readily solved by traditional approaches. Data management consists primarily of the simple storage of information in a way that the relationships among the entities is preserved. Due to the fact that a context can really be composed of any type of data ranging from binary to images, to narratives and audio there is a need for a new model for storage of context data that can handle widely different types of data. Additionally, data management involves the issues of correlations between related types of data. As an example, context C1 may be very similar to context C13-C21 for a given event, thus they should be included in the process of analysis and knowledge creation operations. Related to this idea is that similarity in contexts also is the driving force in the how and what of which contexts are retrieved for a given query.

With the above in mind, there is a need to examine how contextual data might be managed in a previously defined fuzzy data model developed by this author [12]. This model presents an architectural overview of how an original model was developed using fuzzy set theory to manage storage and ambiguous retrieval of information. The elements of the model are presented and how it functions is described. The first part of the next section presents an argument for a new type of paradigm of how data should be thought of, that of the Set model. Problems with this new way of thinking about data organization are then discussed as a beginning for discussion of solutions to the problems of Sets. The section then continues on to discuss a new method of modeling sets that gives the the model an ability to store, manage and retrieve any type of data currently existing and any type of data that may be created in the future. Finally the section presents concepts about how the overlap problems with Sets that create ambiguity in retrieval can now be addressed with new operators based on fuzzy set theory that can identify similarities in data based on time, space and contextual similarity and retrieve the best candidates to satisfy a given query.

Overview of Spatial Data and its Management

Spatial information science is a relatively new and rapidly evolving field. Because global contextual models are highly spatial in many aspects of their operation, including the dimensions of space and time, it is appropriate to look at the issues of context based data management in terms of how spatial data is managed.

Spatial data management systems are an integration of software and hardware tools for the input, analysis, display, and output of spatial data and associated attributes. These systems are being used across a broad range of disciplines for analysis, modeling, prediction, and simulation of spatial phenomena and processes. Applications of spatial data are diverse: natural resource management, traffic control and road building, economic suitability, geophysical exploration, and global climate modeling, to name just a few. In the case of contextual data management, spatial data systems need to extended for a purposes of managing wide types of information such sensed images (i.e., aerial photos, satellite images) and to store data in a number of different raster and vector data structures. A contextual management system based on spatial data management principles may contain digital images, tabular survey data, and text among many other possibilities.

Current spatial data management systems have limitations. One of these is an inability to retrieve and present all the data relevant to a problem to the system user in an orderly and helpful manner. When, for example, a user wants to access information about a particular geospatial region or type of geospatial feature (e.g., tsunami distance and travel information), he or she selects what appears to be an appropriate data entity. This process can be frustrating and error-prone, because, typically, many data entities (maps, photos, etc.) contain information of that type and choosing among them, or even being able to view them all, is very difficult. Furthermore, there may be several data entities (e.g. maps, photos, sensor information) for the same area that have been collected and/or modified at different times for different purposes, and the scales of these maps may differ. Additionally, not all data related to a region may be stored in a database. Some of the data may be in files, scattered across computer systems hard disks.

As an example, a Tsunami has just occurred in the Indian Ocean where map of the coast lines indicate that the area is prone to tsunamis, it has been sensed by NASA from outer space, a ship's captain has noticed a telltale raise in the ocean around his boat and radioed this information to his shipping company and beach vacationers have noticed that the tide has receded dramatically. All of this information is stored somewhere and individually may not have a lot of comprehensive meaning to disaster relief personal in the countries surrounding the Indian Ocean. However as a whole, they clearly indicate a natural disaster with subsequent responses.

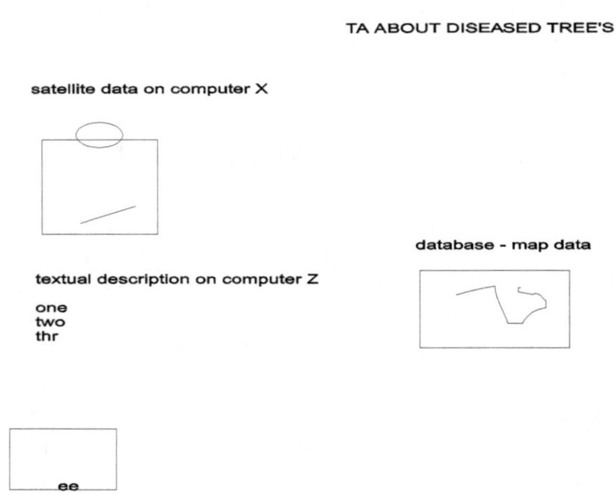

Figure 1. Distributed locations of information that collectively could be stored about a tsunami.

In the above example, a user may know to go to a database and retrieve data about one type of information about tsunamis. However, he or she may be unaware that other types of data not in the database are available that may provide useful and potentially critical information. Even if users are aware of other data, they may not be able to locate.

A goal then of context data management is to develop a way to manage all the type of data that can be found in a context. This can be done by aggregation of context data objects and related contexts into *sets* of data, rather than individual files describing one aspect of the event, in this case a tsunami. This approach can logically associate all related data for a type of event and select the appropriate contextual set based on user-supplied criteria.

The rest of this chapter describes provides an overview of previous work in this area and then a discussion of how contextual information might be stored in the previously defined model.

Context Oriented Data Set Management

Current approaches to data management assume that each individual piece of data, or data file must be managed. For example, in a relational database, as a mountain object would be stored in one row of the table containing mountains. Attributes of the mountain might be stored in a different table. In object-oriented methods, a mountain and its attributes might be stored in a single object. This approach works well in a homogeneous environment where all the data being managed are owned by the application that is managing it. Specifically, the application knows the format of the data it owns and thus manages each and every piece directly. However, this is not practical nor in most cases feasible in a heterogeneous environment that includes a multitude of different applications data. Applications cannot generally read and interpret each other's data. Nevertheless, while data may be for different applications and in different formats, it can still apply to the same geospatial region or problem. When this occurs, there is a basis for the creation of information about relationships among the data, but no mechanism to build the relation because of the differing formats.

This problem can be solved by a shift in the approach to how data is logically thought about and organized. Instead of attempting to manage individual pieces of data, e.g. mountain and attributes of mountains, which may be impossible in a heterogeneous data format environment, one can make the approach less specific, less granular. The key is to manage contextual data on the thematic attributes describing contexts, those of time, space and similarity. In this model specific data objects in a contexts feature vector are not managed they are organized into sets where the set is the lowest level of context data management.

The shift to set management for contextual information produces benefits that address other problems with managing data found in a context feature vector. Specifically, sets can be copies of a base contextual set. These sets can thus be lineages/versions of the base set. Once versions of context sets are established each set can become a particular view of the data included in a context. When views and versions become possible as a result of this approach, then so do multiple information consuming entities with their own lineage trees and domains of control for the sets they define and own. Extending this concept, it is possible to see that multiple views serving multiple users does a very thorough job of addressing the previously user data coupling which is defined to be users modifying their own data and often working on overlapping spatial or temporal themes. Thus the benefits of this approach can have large a impact on a variety of problems. Because the set paradigm is data-format-independent, this is robust approach. The addition of new formats of data that can be described in a context as they are developed, will not cause the new approach to degrade as would current approaches. Instead, one simply adds the new-format data file to the set without any impact to the management and retrieval of such data.

Finally, a set management paradigm can introduce the problem of having multiple members in a set that covers the same geospatial region. This is referred to as ambiguity and was addressed through the application of fuzzy set theory to the metadata that manages the set abstraction. Fuzzy set theory can be used to make a generalized comment about the degree of possible membership a particular data file might have in a set covering a specific geographic region.

Contextual Set Ambiguity

Dataset ambiguity in contexts refers to the fact that for a given query or selection it may be impossible to select and exact match for the query because multiple sets of context may satisfy the query fully or partially. For example, if a query is interested in all data about an event located at a geographic point on the ground, multiple sets may have overlapping boundaries that the point can fall inside, thus the question becomes which set to return for a query. Another example can be found when one considers the spatial data in a context where the boundaries of objects are approximately known but not precisely known. For example if one is mapping the extent of the spreading wave of a tsunami, the edge of the wave and thus its boundary may be one meter wide or it may be considered to be hundreds of meters wide. The selection of information about the edge of the tsunami wave then becomes an ambigious problem. Because multiple contextual sets about a given tsunamis boundary may exist, perhaps one defines the edge to be one meter and the other for the same geographic location defines the edge to be 100 meters the question is which context sets data should be retrieved for a query. Because context have multiple dimensions, this problem can also exist for the temporal dimension of contextual sets and the similarity dimension. It also may exist for the impact dimension.

Ambiguity in contextual data sets impacts their use in a fairly significant fashion and has been studied for spatial data but not the additional dimensions of context set data. Contextual data sets (CDS) could be organized into large databases. Users of the database could then create spatial or geographic queries to the database to retrieve CDS data that is of interest to them based on geographic extent. This process of doing this is sometimes referred to as geographic information retrieval if the queries are for spatial data (GIR) [19]. GIR seeks to deal with spatial uncertainty and approximation in the methods by which traditional spatial data is indexed and retrieved.

A key shift in the new model for CDS management is towards being less granular in the management of CDS data. Instead of managing geographic entities such as one might find in a GIR database, or for that matter the dimensional entities of temporality and similarity, the smallest unit of management in this approach is a single covering logical device that of a set for CDS data. This shift to being less granular has a variety of benefits, but it can introduce further ambiguity in selecting and defining sets with multiple overlapping coverage's. In this sense a coverage is the data found in a CDS that describes the dimension of space, time, similarity and impact. With this in mind, the new model defines ambiguity as condition where multiple tracts of CDS data may satisfy a given query for a particular geographic location, point in time, type of similarity or type of impact. When this is the case, the question becomes which set of data should be returned for a spatial query to retrieve the correct coverage?

To illustrate this point, consider the case where two contextual datasets have a coverage that contains the origin of a tsunami. One dataset contains is satellite imagery and the other is sensor information from the ocean. This is a "point in polygon" type of ambiguity problem. The center of the Tsunami is the point and a polygon is the rectangular bounding polygon of each spatial coverage. The expected ambiguity between trying to select between these coverage can be seen in Figure 1.

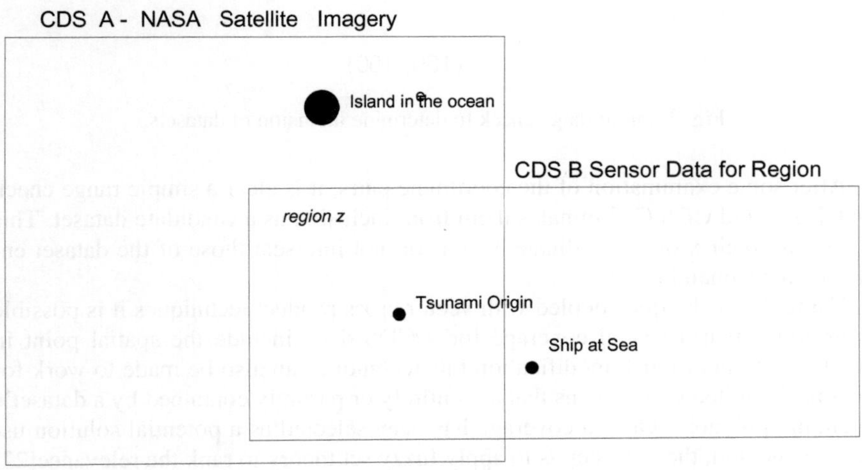

Fig. 1. Example of data set ambiguities for spatial coverage of the origin of a tsunami.

In the above example, the question is whether one wants CDS B or CDS A for a query about tsunami CDS data at time T_0. This is an example of ambiguous spatial data, and a point in polygon ambiguous problem.

Ultimately, the choice of which set to choose in an ambiguous problem should be left up to the analyst, or the person using the data. However, application of fuzzy set theory and computational geometry can be applied to presenting potential datasets in ways that might solve the problem shown above. The solution involves a stepwise algorithm in finding a solution.

First must be identified the CDS datasets that potentially might solve the query for data about the tsunami. Step one would then be to do a simple range check to see if the location of the tsunamis' y coordinates are within the y extent of any dataset known to the system. An example of how this could be accomplished is illustrated in Figure 2.

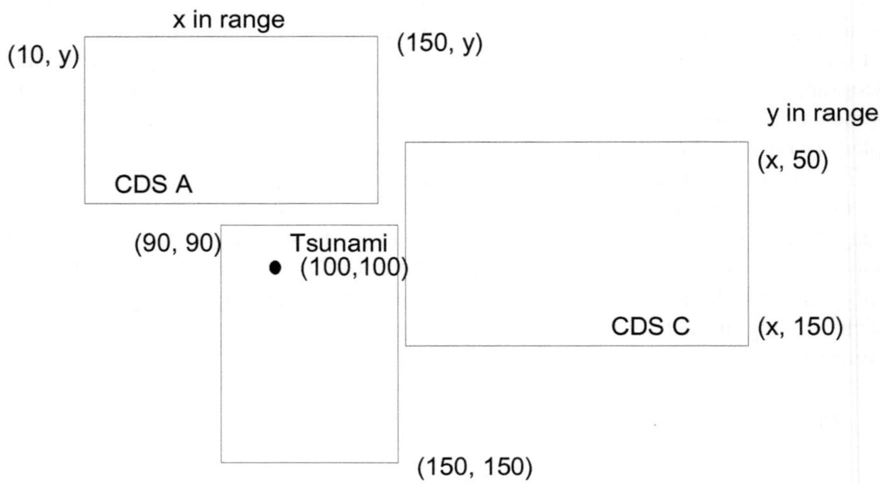

Fig. 2. Initial range check to determine inclusion of datasets.

After some examination of the coordinate pairs, it is clear a simple range check on CDS A and CDS C eliminates them from inclusion as a candidate dataset. This is because their x or y coordinate extents do not intersect those of the dataset enclosing the tsunami point.

Using this technique coupled with vector cross product techniques it is possible to establish that a spatial coverage for a CDS does include the spatial point in question. Without much modification this technique can also be made to work for regions delimited by polygons that are entirely or partially contained by a dataset's bounding polygon. Once a coverage has been selected as a potential solution using this method, the next step is to apply fuzzy set theory to rank the relevance[22] of the coverage to the point of origin of the tsunami.

The approach used in this model is to do this in one of several fashions. The first of these would be to calculate the distance between the tsunami's point of origin point and the centroid of a bounding polygon. The distance value could become a component in the return value for the fuzzy membership function for the spatial aspects of CDS data, MSpatial() which is discussed later. In this case, the smaller the distance, the more centrally located the point representing the tsunami's point of origin is to the coverage being considered. This scenario is shown in figure 4.

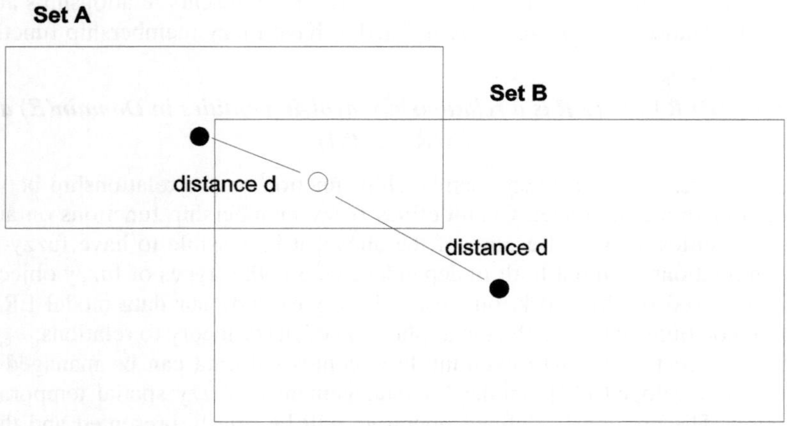

Fig. 3. Application of distance as a return value for a fuzzy function.

The value of the fuzzy membership function MSpatial() to select for Set B may be .9, whereas the value of the fuzzy function MSpatial() for Set A might be .3. Higher values of MSpatial() would reflect the fact that the distance to the nearest centroid was smaller. This would then suggest that because the point represented by the tsunami's origin is more centrally located in Set A, that this set is a much better set to use if one wants to examine data for Pullman and the surrounding area.

While this approach makes sense, it is simplistic. Therefore, weighting factors in conjunction with attributes of the data sets themselves might also be considered. An example of this might be that both sets have metadata for "data accuracy", and the "time last data collection". In a CDS model one might argue that the data accuracy is very important (.9) because improved accuracy would be expected to reflect current reality better. Following this logic, a scheme then might weight "time of last data collection" very heavily. We might give lesser weight to the accuracy weight (.3).

A weighting scheme to use in retrieval might develop in the CDS model a function to assist in resolving data ambiguity in this case might become

$$\mathbf{C_{weight}() = distance * (.9 * days\ since\ last\ edit + .3 * time\ of\ day\ edited)}$$

In above scheme, geometric properties can be combined with attribute properties for a CDS set to solve an ambiguous selection and retrieval problems.

Rationale For Fuzzy ERD's to Manage Contextual Data

Chen [8] defines an Extended Entity Relation (EER) model to consist of the triple

$$M = (E, R, A)$$

where M represents model, E represents entities, R represents relationships and A represents relationships. E, R, A are defined to have fuzzy membership functions. In particular:

R = {U_r (R)/R | where R is a relationship involving entities in Domain(E) and Ur(R) $\in$ [0,1]

In this case, ***Ur()*** is a fuzzy membership function on the relationship between two entities in a data model. Chen defines fuzzy membership functions on attributes and entities as well. Because of the above, it is possible to have fuzzy relations on relations, without built in dependencies on other types of fuzzy objects in a model. Based on this work, our research now extends our data model ERD to defining notations that describe the application of fuzzy theory to relations.

The next section we will examine how contextual data can be managed in a previously developed [12] model for management of fuzzy spatial temporal information. The previously defined operators will be briefly presented and then a discussion will be made about how the model supports or does not support that of contextual data management.

A Fuzzy ERD Model for Context Management

The data model in figure 2 provides an initial foundation to address problems inherent with management of context data. The data model was developed to model spatial and temporal information which are two key dimensions of contextual data. One of the problems with contexts, as with spatial and temporal data, is that of ambiguity in the selection and retrieval of data. These problems can be addressed by the application of fuzzy set theory. For example, several overlapping coverages for tsunami information could exist based on time and space. In this sense overlapping coverage is defined to be multiple contexts with information fully or partially about the same event, e.g. the origin of the tsunami. Keep in mind there is a tendency to think of such information as geo-spatial but it may also include images and textual descriptions. The key concept in retrieval is to find the most "appropriate" coverage for a given query. Appropriate is a term that can only be defined for the consumer of the information, the user. The logical question becomes which context data set to select and use for a given purpose.

Overlapping contextual spatial coverages are a type of ambiguity. We have also identified that there can be overlapping contextual temporal locations. We can also have overlap in the similarity of contexts and their impact dimension which are not addressed in this chapter. When considering the problem of overlap in selection and retrieval of information the types of overlap that can be present must also be considered. Overlap can be partial or complete overlap with different descriptive characteristics to coverage such as different projections, scale and data types. These can also become complications to ambiguity of selection. Considering this situation, it is clear that ambiguity on an attribute of spatial data can compound with other ambiguities about the same data. This can have the potential of leading to much larger ambiguities.

To date, a lot of work has been done in the development of fuzzy set theory, and techniques for decision making such as using Open Weighted Operators [4]. Little of this work has been applied to the management of contextual data. In particular, theoretical discussions needs some form of implementation to solve real world problems. What is needed is the application of theory and a representational notation. The application could then be used to solve real world problems such as data ambiguities. The figure 2 data model was extended with new types of fuzzy operators that address ambiguous selection problems to create a more powerful model that can deal with the problem of ambiguous data.

Contextual Subsets

The first new notational convention is the context subset symbol. The *Subset* symbol defines a new type of relationship on an entity, that is it borrows from object oriented constructs, that of the "bag". An entity with the subset symbol defined on one of its relations is a non-unique entity, unlike most entities in an ERD model. The rationale for its existence is that multiple copies a *Subset* containing the same elements can exist for different overlapping temporal, spatial, impact and similarity coverages for a given event, e.g. the tsunami. This circumstance can occur as a result of various versions of the same *Subset*, or normal editing operations. The symbol is defined as:

$$\subseteq$$

By its nature of being a non-unique entity, a relationship with the *Subset* definition, also is a fuzzy relationship. This is due to the fact that when one desires to view a *Subset*, the question becomes which one should be selected. Because *Subsets* are discrete, the *Subset* symbol occurs in our model with the symbol for fuzzy relation M() which is defined next.

Fuzzy Relation M()

Fuzzy theory literature [7] defines a membership function that operates on discrete objects. This function is defined as M() and has the following property:

$$\text{Similiar}(a) = \begin{cases} 1 & \mid \text{if } a \in \text{domain}(A) \\ 0 & \mid \text{if } a \notin \text{domain}(A) \\ [0,1] & \mid \text{if } a \text{ is a partial member of domain}(A) \end{cases}$$

This function is particularly useful in contexts where overlapping coverages of the same event space may exist, but some coverage for a variety of reasons may be more relevant to a particular concept such as a desire to perform editing of surrounding regions. The actual definition of how partial membership if calculated has been the subject of much research including the application of Open Weighted Operators (OWA) [3,10] and the calculation of relevance to a concept [9].

The data model developed for contexts model in this chapter seeks to provide alternative view support for overlapping geospatial coverages. Because of the ambiguities induced by this, we introduce a notation that represents the fuzzy relation resolved by the definition of the function M(). This symbol is referred to as the fuzzy relation M() symbol and may be displayed in and ERD model along the relations between entities. It has the following notation:

$$\sum$$

This symbol makes no comment about the nature or calculation of M() per se, but does suggest that the M() function is evaluated when a query on the relationship in the ERD is generated. The query returns a ranked set of items with a similarity value in the range of [0,,1]

Another property of the function M() is that it reflects the fuzzy degree of relation that entities have with other entities.

Fuzzy Directionality

Fuzziness in the context data model is not bi-directional on a given relationship. Therefore there needs to be some indication of the direction fuzziness applies. This is denoted by the inclusion of the following arrow symbols on the fuzzy relationship. These arrows are found to the left of the fuzzy symbol and point in the direction that the fuzzy function M() or MSpatial() applies. If a fuzzy relationship is defined in both directions, which implies a type of m:n relation, the symbol is a double headed arrow.

Directional fuzziness for the M() or MSpatial() function, points in the direction the function is applied for selection and is denoted by the following symbols:

$$\leftarrow, \uparrow, \rightarrow, \downarrow$$

Bi-directional application, is a member of the class of m:n relations and is denoted by:

$$\leftarrow\rightarrow$$

Discretinizing Function D()

Because of the data ambiguities mentioned previously, the new context based data model uses time as an attribute in describing data. However, this leads to temporal ambiguities in the selection of a *Subset* of data because the *Subset* can exist at many points in time. However, there are certain points in time where the relevance of data to a concept or operation, e.g. a selection, query is more relevant. Therefore the relation of *Subset* to *Temporal Location* entity can have fuzzy logic applied. Time is not a discrete value, it is continuous, and therefore it is referred to as a continuous field. Some attempts have been made to discretinize continuous temporal data by Shekar [9] using the *discretized by* relation. But no known attempts have been made to deal with this in a fuzzy fashion This leads to the need to define a new function D() that can be used to calculate discrete fuzzy membership value over continuous fields.

In the new contextual model for data management, the inclusion of continuous field data is useful. This is due to the fact that sets of data not only cover a geographic extent, but they also cover this particular extent for a period of time and then can be replaced by another set, perhaps not of the same geographic coverage but at a different point in time.

If time is non-discrete and a function must be developed, the question becomes how to represent continuous data in a fashion that a function can make computations on the data and return a discrete value representing membership. Upon examination of this issue, non-discrete data can be defined as a bounded range [m,n] where the beginning of the continuous temporal data starts at time m and terminates at time n. This representation then makes it possible to develop the function D() and its behavior over continuous data.

In this function one wants to think of a window of time that a set of context data was created at time $t_{m,}$ spanning to a point in time where the data is no longer modified, t_n above equation, the range $[t_m,t_n]$ is referred to as the "window" because it is a sliding window on the continuous field that one seeks to determine the degree of membership of selection point of time t to be. A function that can then be used to retrieve relevant sets of contexts can be defined as:

$$\mathrm{INRANGE}([t_m,t_n], t) = \{\ \mathrm{ABS}\ [\ (t - t_m)\ /\ (t_n - t_m)\ ]\ \}$$

where ABS() is simply the absolute value of the calculation.

The effect of D() is to make a discrete statement about non-discrete data, which makes it possible to make assertions about fuzziness and possibilities. The statement is of course relative to the bounded range [m,n] and therefore D() should be formally denoted as:

$$\mathrm{MSpatial()}_{[m,,n]}$$

when referring to value returned for particular calculation of the function MSpatial()

The application of MSpatial() is found in the dataset management model on the fuzzy notation denoted by the symbol :

This symbol is displayed on the model oriented such that relation lines intersect the vertical lines of the symbol. This notation means that the relation is fuzzy and is determined by the discretinizing function MSpatial() as defined above. For the purposes of the data management model, discretinizing functions are applied to temporal entities that define a *Subset* of sets of contextual data by a temporal location. They can however, be applied to any type of continuous field data.

Fuzzy Relation MSpatial()

The function M() is not a complete function for the solution to selection problems in the developed ERD model. This is because it does not consider the centrality of a point P composed of an x,y and perhaps z component that one wishes to retrieve context data about. This lead to the creation of of a function referred to as MSpatial(). The characteristic function MSpatial() needs to contain a function that measures distance, a new term, d, that can be derived in the following manner:

$$d = \min\left(\sqrt{(centroidx - po\text{int}\, x)^2 + (centroidy - po\text{int}\, y)^2}\right)$$

$$centroidx = .5 * (s1x2 - s1x1)$$

$$centroidy = .5 * (s2y2 - s2y1)$$

The above equation refers to a rectangular bounding hull created around a geographic coverage. Centroid$_n$ is the centroid of the bounding hull found by finding the mid point of of side one for centroidx and the mid point of side 2 in y for centroidy.

Pointx and pointy are the coordinates of a spatial entity or center of a region of interest that one is seeking the most centrally located coverage for. The d value in the a characteristic function for MSpatial() then becomes a measure of the minimum distance of a coverage's centroid to a spatial entity. The effect of the equation is to find a contexts coverage that is most central to the spatial entity of interest. The goal is to weight the characteristic functions values with a measured degree of centralization to a spatial center of interest when selecting fuzzy data for a particular problem.

The application of MSpatial() is found in the dataset management model on the fuzzy notation denoted by the symbol :

$$\sum_{S}$$

This symbol is displayed on the model oriented such that relation lines intersect the vertical lines of the symbol. This notation means that the relation is fuzzy and is determined by the MSpatial() function

Extended Data Model for The Storage of Context Data Sets

With an understanding of the issues found in retrieval of context data sets and some new fuzzy characteristic functions and notations a new data model can be presented for management of sets of context data. This model considers the vagaries of ambiguity in time and space selection which are dimensions of contexts but does not support the contextual dimensions of similarity and impact. The model is presented in figure 2.

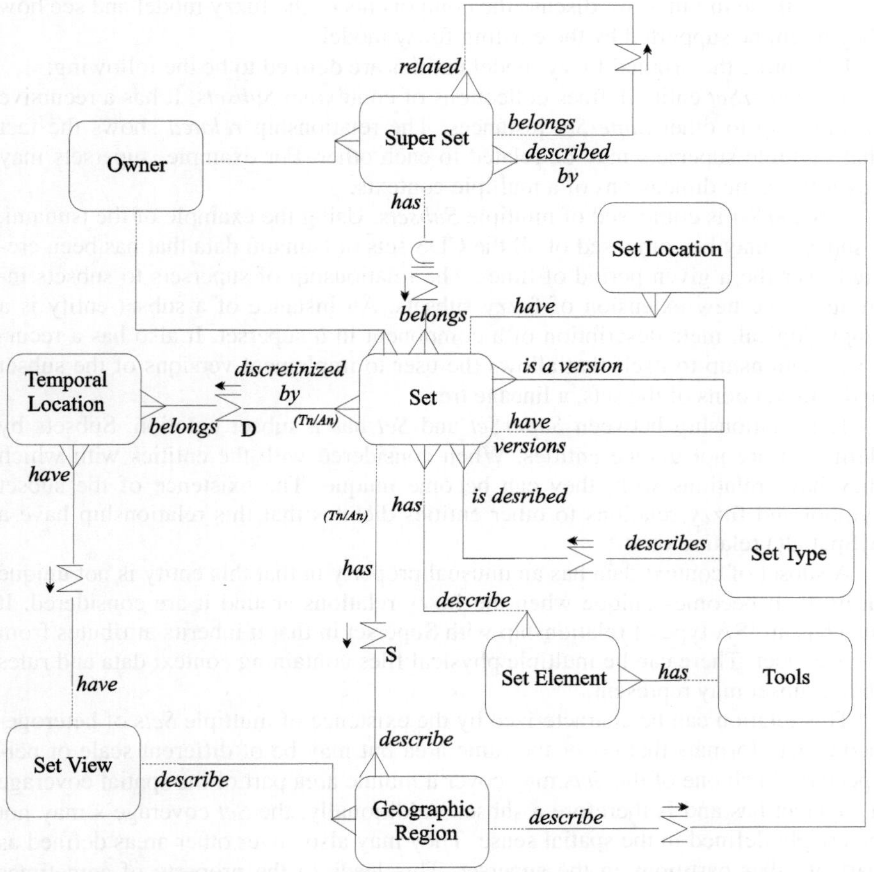

Figure 2. Extended fuzzy model that can be applied to context set management.

We now examine some characteristics of context data for how the above model for management of spatial data will support these. Contexts have the following properties:

- four dimensions that uniquely characterize them, time, space, impact and similarity
- do not have an owner because they stream from their sources
- do not have a specific location they reside because they float around the internet
- do not have a particular type associated with them because contexts can be composed of many different data types
- might have specific tools that process them depending on the consumer of the information
- do not have views of the data that are currently defined
- super contexts are composed of contexts which are composed of elements

With these in mind, we discuss the components of the fuzzy model and see how they might be supported by the existing fuzzy model.

In figure 2 the original fuzzy model entities are defined to be the following:

The *SuperSet* entity defines collections of contextual *Subset*s. It has a recursive relationship to other *SuperSet* instances. The relationship *related* shows the fact that multiple supersets may be related to each other. For example, supersets may cover the same dimensions of a multiple contexts.

A *SuperSet* is composed of multiple *Subset*s. Using the example of the tsunami, a superset may be composed of all the CDS sets of tsunami data that has been created over the a given period of time.. The relationship of supersets to subsets introduces the new extension of fuzzy subsets. An instance of a subset entity is a single logical, meta description of a component in a superset. It also has a recursive relationship to itself that allows the user to implement versions of the subset and thus versions of the sets, a lineage tree.

The relationship between *SuperSet* and *Set* has a subset notation. Subsets by definition are not unique entities. When considered with the entities with which they have relations with, they can become unique. The existence of the subset symbol and fuzzy relations to other entities dictates that this relationship have a MSpatial() relation on it.

A subset of context data has an unusual property in that this entity is not unique in itself. It becomes unique when the fuzzy relations around it are considered. It also has an ISA type of relationship with Superset in that it inherits attributes from the superset. There can be multiple physical files containing context data and rules that a subset may represent.

This relation can be characterized by the existence of multiple *Set*s of heterogeneous data formats that cover the same area but may be of different scale or perspective. Each one of the *Set*s may cover a minute area part of the spatial coverage a superset has and is therefore a subset. Additionally, the *Set* coverage's may not be crisply defined in the spatial sense. They may also cover other areas defined as part of other partitions in the superset. This leads to the property of sometimes being unique and sometimes not being unique.

The *Set* and *Super Set* entities and their subset relationship support the concepts of contexts. Specially, they support the idea that super contexts are composed of contexts and a feature vectors data in a given context is composed of elements also referred to as attributes in the context paradigm.

The *Temporal Location* entity represents a locational time definition for a data set. It has temporal attribute values and geospatial coordinates that collectively create identifiers for a particular data set.

The D() relation between *Temporal Location* relation and *Set* can be used to select context *Set*s that cover a given range in time. These can occur due to the existence of long editing transactions on the data. A *Set* is not instantly updated during a long transaction. Certain points in the existence of the *Set* may be more of interest when selecting a set to view, but all are valid descriptions of the subset. The D() function exists as a sliding window of possibility for selecting a *Set* that has existed and was updated over a period of time. This allows one to select the *Set* in a specific time range.

The *Temporal Location* entity exists because there is a need for given data sets to map to various locations in time and spatial coverages. The relationship "discretized" was originally defined to map a value to a continuous field. In this case the discretized relation has been extended to represent a discretized function where the continuous field is time. Because this function can relate a data set to various points in time and coverage of several different spaces, this function is a fuzzy function, D() that selects on time and spatial definitions for a *Set*.

In this case *Temporal Location* supports the concepts that contexts have a dimension of time that describes them. The aggregation of this fact as it relates to specific contexts then defines a super contexts window of temporal existence embodied in the *Super Set* Entity. The D() operator reflects the fact that contexts stream data as they are created, thus there is a need to select data that may span ambiguous moments in time.

The *Geographic Region* entity locates a *Subset* of data by the type of spatial coverage it has. This entity works in conjunction with the *Temporal Location* entity to locate a set of data in time and space. The rationale is that for a given spatial area, there may be multiple coverages generated over time. Therefore, the problem becomes one of locating spatial data in 2D space.

The *Geographic Region* entity has a M() relationship with *Set*. The rationale is that because context *Set*s of data can be overlapping in spatial coverage for a given point in space, selection of a subset becomes an ambiguous problem. The MSpatial() symbol then implies that selection of *Set*s covering a geospatial point needs to done using some type of fuzzy selection.

The *Geographic Region* entity is not clearly defined in the context model at the present because the regions that contexts are created for is assumed to be fixed and thus is not ambiguous. However, it can be logically argued that contexts may not be registered exactly over the same geographic point. This could be the subject of future investigation.

The *Set Type* entity describes the type of data a *Subset* may contain. An example of the expected types where "image", "raster", etc. This entity was also a candidate for fuzzy notation extension, following this section. The *Set Type* entity is not defined in the context model because a set maps to a feature vector and a feature vector is composed of multiple types of disparate data that are stored at the media location described by *Set Location*.

The *Set View* entity in the model provides a repository for information about various views of data that may exist for a given *Subset. Set View* makes it possible to have multiple views of the same data set. Such views would differ by such things as a datasets perspective, scale or projection. The *Set View* entity is not supported in the context model and therefore there is no current mapping or application of its functionality.

The *Set Location* entity describes the physical location of the *Subset* and thus a contexts data. This entity is required because of the need for a model where data can be distributed around a computer network or around the internet. This entity provides the potential to support distributed repository mechanisms because parts of the database are not in the same physical data space at all times. It also provides a way to have alternative views of data that are not centrally located. This entity is very highly supported in the context model where contexts are hyper distributed around the internet. The context model will probably spend considerably more time developing the concepts behind *Set Location*.

Analysis of Coverage Support of the Fuzzy ERD for Contextual Data Sets

Having established that the Fuzzy ERD model does support management of Contextual data sets, it is useful to determine how much of the model is extra and could be trimmed to refine the model.

The above section finds that the entities that are not utilized when mapping the context model onto the fuzzy storage model are *Owner, Set View*, and *Set Type*. This is due to the characteristics of contextual data discussed previously that differ from geo-spatial data. Entities that are marginally supported are *Tools* and *Geographic Location.* If we determine a coverage ratio for the context models mapping onto the fuzzy set management model it come to the following:

2 entities partially supported which arbitrarily are counted as 1
3 entities that do not map to the context model
6 entities that fully support the model

This produces a coverage ratio of 1+ 6 / 11 = 63%. This means 37% of the previously developed fuzzy ERD model is not utilized and potential for elimination in a new tailored model. This ratio could be increased if the ERD entities that are partially supported by the mapping were honed to be fully functional in the support of the context model. If this is done, the ratio of utilized entities to non utilized entities become 8/11 or 72%. This means that 28% of the existing model does not really support contexts and is a candidate for removal.

Future Research

This research merges the concepts of global contexts with that of an existing fuzzy data model for management of spatial and temporal information. What is discovered in the process is that the dimensional elements of contexts, those of time and space lend themselves well to management by such a model. The dimensions of similarity and impact are not supported in such a model and thus a subject of future research. The final results of the analysis of coverage suggest that a new type of data model should be developed that is more tuned to support of the contextual model. Additionally, the performance impact of the fuzzy operators should be also evaluated to see how sharply they affect retrieval of information. Mechanisms should also be examined that can help the fuzzy selection functions adapt to their own performance in such a way that feedback can improve their performance. Much work remains to be done in this newly emerging area of a new paradigm for information sharing on a global scale.

References

[1] Rosemann, M., Recker, J.: Context-aware process design: Exploring the extrinsic drivers for process flexibility. In: Latour, T., Petit, M. (eds.) 18th international conference on advanced information systems engineering. Proceedings of workshops and doctoral consortium, pp. 149–158. Namur University Press, Luxembourg (2006)

[2] Schilit, B.N.A., Want, R.: Context-aware computing applications" (PDF). In: IEEE Workshop on Mobile Computing Systems and Applications (WMCSA 1994), Santa Cruz, CA, US, pp. 89–101 (1994)

[3] Schilit, B.N., Theimer, M.M.: Disseminating Active Map Information to Mobile Hosts. IEEE Network 8(5), 22–32 (1994)

[4] Dey, A.K.: Understanding and Using Context. Personal Ubiquitous Computing 5(1), 4–7 (2001)

[5] Bolchini, C., Curino, C.A., Quintarelli, E., Schreiber, F.A., Tanca, L.: A data-oriented survey of context models (PDF). SIGMOD Rec. (ACM) 36(4), 19–26 (2007), http://carlo.curino.us/documents/curino-context2007-survey.pdf

[6] Schmidt, A., Aidoo, K.A., Takaluoma, A., Tuomela, U., Van Laerhoven, K., Van de Velde, W.: Advanced Interaction in Context (PDF). In: 1th International Symposium (1999)

[7] Burrough, P.: Natural Objects With Indeterminate Boundaries. Geographic Objects with Indeterminate Boundaries. Taylor and Francis, Abington (1996)

[8] Chen, G., Kerre, E.: Extending ER/EER Concepts Towards Fuzzy Conceptual Data Modeling, MIS Division, School of Economics & Management. Tsinghua University, Bejing, China

[9] Morris, A., Petry, F., Cobb, M.: Incorporating Spatial Data into the Fuzzy Object Oriented Data Model. In: Proceedings Seventh International Conference IPMU (1998)

[10] Shekar, S., Coyle, M., Goyal, B., Liu, D., Sarkar, S.: Data Models in Geographic Information Systems. Communications of the ACM 40 (April 1997)

[11] Yager, R., Kacprzyk, J.: The Weighted Averaging Operators, Theory and Applications. Kluwer Academic Publishers, Boston (1997)

[12] Vert, G., Stock, M., Morris, A.: Extending ERD modeling notation to fuzzy management of GIS datasets. Data and Knowledge Engineering 40, 163–169 (2002)

[13] Larson, R.: Geographic Information Retrieval and Spatial Browsing. School of Library and Information Sciences, University of California, Berkely (1999)

[14] Morris, A., Petry, F., Cobb, M.: Incorporating Spatial Data into the Fuzzy Object Oriented Data Model. In: Proceedings of Seventh International Conference IPMU (1998)

Repercussions of Fuzzy Databases Migration on Programs

Mohamed Ali Ben Hassine, José Galindo, and Habib Ounelli

Abstract. Fuzzy databases have been introduced to deal with uncertain or incomplete information in many applications demonstrating the efficiency of processing fuzzy queries. For these reasons, many organizations aim to integrate the fuzzy databases advantages (flexible querying, handling imprecise data, fuzzy data mining, ...), minimizing the transformation costs. The best solution is to offer a smoothly migration toward this technology. However, the migration of applications or databases in enterprises arises from changes in business demands or technology challenges. The need for this migration is to improve operational efficiency or to manage risk, data migration outage, as well as performance. This chapter is about the migration towards fuzzy databases. We present our migration approach and we concentrate on their repercussions on programs.

1 Introduction

Legacy information systems are typically the backbone of an organization's information flow and the main vehicle for consolidating business information. They are thus critical missions, and their failure can have a serious impact on business. Therefore, these legacy systems are requested to be flexible and efficient to cope with

Mohamed Ali Ben Hassine
Department of Computer Sciences, Faculty of Sciences of Tunis, El Manar 1,
Campus Universitaire, 2092, Tunis, Tunisia
e-mail: mohamedali.benhassine@fst.rnu.tn

José Galindo
Department of languages and Computer Sciences, University of Malaga, 29071, Spain
e-mail: jgg@lcc.uma.es

Habib Ounelli
Department of Computer Sciences, Faculty of Sciences of Tunis, El Manar 1,
Campus Universitaire, 2092, Tunis, Tunisia
e-mail: habib.ounelli@fst.rnu.tn

J. Kacprzyk et al. (Eds.): Uncer. Appro. for Spatial Data Model. and Process., SCI 271, pp. 175–193.
springerlink.com

rapidly changing business environments and advancement of services. The need for migration of applications or databases in enterprises arises from changes in business demands or technology challenges either to improve operational efficiency or to manage risk, data migration outage, as well as performance. In reality, the risk of failure is usually too great for organizations to seriously contemplate a migration approach. Another very real concern stems from the fact that technology and business requirements are constantly changing. Thus, at the end of a long process, an organization might find itself with a redeveloped system based on obsolete technology that no longer meets its business needs.

This need is to preserve established business rules and practices in the old system, and to manage the transition of valuable human resources locked in maintaining legacy to more modern information systems.

Closer to our context and according to Bellman and Zadeh [3], "much of the decision making in the real world takes place in an environment in which the goals, the constraints, and the consequences of possible actions are not known precisely". Management often makes decisions based on incomplete, vague, or uncertain information. In our context, the data which are processed by the application system and accumulated over the lifetime of the system may be inconsistent and may not express the reality. In fact, one of the features of human reasoning is that it may use imprecise or incomplete information and in the real world, there exists a lot of this kind of data. Hence, we can assert that in our everyday life we use several linguistic terms to express abstract concepts such as young, old, cold, hot, and so forth. Therefore, human-computer interfaces should be able to understand fuzzy information, which is very usual in many human applications. However, the majority of existing information systems deal with crisp data through crisp database systems [14]. In this scenario, fuzzy theory has been identified as a successful technique for modelling such imprecise data and also for effective data retrieval. Accordingly, fuzzy databases (FDBs) have been introduced to deal with uncertain or incomplete information in many applications demonstrating the efficiency of processing fuzzy queries even in classical or regular databases. Besides, FDBs allow storing fuzzy values, and of course, they should allow fuzzy queries using fuzzy or non fuzzy data [8, 17].

Facing this situation, many organizations aim to integrate flexible querying to handle imprecise data or to use fuzzy data mining tools [15], minimizing the transformation costs. A solution for the existing (old) systems is the migration, i.e., moving the applications and the databases to a new platform and technologies. Migration of old systems, or legacy systems, may be an expensive and complex process. It allows legacy systems to be moved to new environments with the new business requirements, while retaining functionality and data of the original legacy systems. In our context, the migration towards FDBs does not only constitute the adoption of a new technology, but also, and especially, the adoption of a new paradigm. Consequently, it constitutes a new culture of development of information systems. However, with important amounts invested in the development of relational systems, in the enrollment and the formation of "traditional" programmers, and so forth, enterprises appear reticent to invest important sums in the mastery of a new fuzzy paradigm. Therefore, we have proposed a migration approach [4, 6] toward this technology, allowing them to keep the existing data, schemas, and applications, while integrating the different fuzzy concepts to benefit of the fuzzy information processing. It will reduce the costs of the transformations and will encourage the enterprises to adapt the concept of fuzzy relational database (FRDB).

This chapter focuses on the repercussions of fuzzy migration on applications programs. We propose some methods to adapt the application programs of the legacy systems to new fuzzy data: wrapping, maintenance, redevelopment, and migration through rewriting access statements. These methods are related strongly to the strategies of FRDB migration presented in our previous work [4]. First, we present a very brief overview about FRDB. Second, we present our three migration strategies. Third, we present the methods which we propose to migrate legacy programs in order to treat new fuzzy data. Fourth, we illustrate these methods with an example. Finally, we outline some conclusions and suggest some future research lines.

1.1 Introduction to Fuzzy Relational Databases

A FRDB is as an extension of a RDB. This extension introduces fuzzy predicates or fuzzy conditions under shapes of linguistic expressions that, in flexible querying, permits to have a range of answers (each one with its membership degree) in order to offer to the user all intermediate variations between the completely satisfactory answers and those completely dissatisfactory [9, 17, 35]. Yoshikane Takahashi [29] defined FRDB as "an enhanced RDB that allows fuzzy attribute values and fuzzy truth values; both of these are expressed as fuzzy sets". Summarizing, this extension introduces fuzzy information processing (fuzzy attributes, fuzzy queries, fuzzy data mining, ...). A good reference about new trends in FDBs and a good definition of the main terms is in [16]. One of the main applications, fuzzy queries, includes fuzzy comparators, fuzzy expressions, fuzzy conditions, and fulfillment degrees, which permits to rank the answers. A good review about fuzzy query systems may be found in [35]. There are many forms of adding flexibility in FDBs. The simplest technique is to add a fuzzy membership degree to each record, an attribute in the range [0,1]. However, there are others kind of databases allowing fuzzy values, using fuzzy sets, possibility distributions, fuzzy degrees associated to some attributes and with different meanings (membership degree, importance degree, fulfillment degree...). The main models are those of Prade-Testemale [27]), Umano-Fukami [31, 32], Buckles-Petry [10], Zemankova-Kaendel [36] and GEFRED by Medina-Pons-Vila [26]. Recently, some approaches exist about fuzzy object-oriented databases, like [2].

This chapter deals mainly, with the GEFRED model [26], and some later extensions [17]. This model constitutes an eclectic synthesis of the various models published so far with the aim of dealing with the problem of representation and treatment of fuzzy information. One of the major advantages of this model is that it consists of a general abstraction that allows for the use of various approaches, regardless of how different they might look. In fact, it is based on the generalized fuzzy domain and the generalized fuzzy relation, which include respectively classic domains and classic relations. These authors also include a fuzzy modeling tool

(FuzzyEER), the fuzzy language FSQL (a fuzzy extension of SQL to cope these topics, specially the fuzzy queries), FIRST-2 (definitions in order to implement a real FDB on a classical DBMS), and some applications. Inside FIRST-2, we can find the Fuzzy Metaknowledge Base (FMB), i.e., the data dictionary or catalog which represents the necessary information related to the imprecise nature of the new collection of data processing (fuzzy attributes, their type, their objects such as labels, quantifiers, etc.).

1.1.1 Fuzzy Attributes

In order to model fuzzy attributes, we distinguish between two classes of fuzzy attributes: fuzzy attributes whose fuzzy values are fuzzy sets (or possibility distributions) and fuzzy attributes whose values are fuzzy degrees. Each class includes some different fuzzy data type [17].

Fuzzy Sets as Fuzzy Values: These fuzzy attributes may be classified in four data types. In all of them, the values Unknown, Undefined, and Null are included:

- Fuzzy Attributes Type 1 (FTYPE1): These are attributes with "precise data", classic or crisp (traditional with no imprecision). However, we can define linguistic labels over them, and we can use them in fuzzy queries. This type of attribute is represented in the same way as precise data, but they can be transformed or manipulated using fuzzy conditions. This type is useful for extending a traditional database, allowing fuzzy queries to be made about classic data. For example, enquiries of the kind "Give me employees that earn a lot more than the minimum salary".
- Fuzzy Attributes Type 2 (FTYPE2): These are imprecise data over an ordered referential. They admit both crisp and fuzzy data, in the form of possibility distributions over an underlying ordered dominion (fuzzy sets), such as "he is approximately 2 meters tall". For the sake of simplicity, the most complex of these fuzzy sets are supposed to be trapezoidal functions (Figure 1).

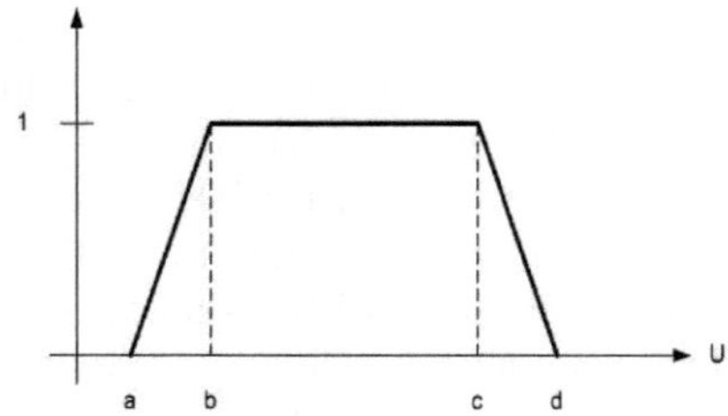

Fig. 1. Trapezoidal, linear and normalized possibility distribution.

- Fuzzy Attributes Type 3 (FTYPE3): They are attributes over "data of discreet non-ordered dominion with analogy". In these attributes, some labels are defined (e.g., "blond", "red", "brown", etc.) that are scalars with a similarity (or proximity) relationship defined over them, so that this relationship indicates to what extent each pair of labels resemble each other. They also allow possibility distributions (or fuzzy sets) over this dominion, for example, the value (1/dark, 0.4/brown), which expresses that a certain person is more likely to be dark than brown-haired. Note that the underlying domain of these fuzzy sets is the set of labels, and this set is non-ordered.

- Fuzzy Attributes Type 4 (FTYPE4): These attributes are defined in the same way as Type 3 attributes without it being necessary for a similarity relationship to exist between the labels.

Fuzzy Degrees as Fuzzy Values: The domain of these degrees is the interval [0,1], although other values are also permitted, such as a possibility distribution (usually over this unit interval). The meaning of these degrees is varied and depends on their use. The processing of the data will be different depending on the meaning. The most important possible meanings of the degrees used by some authors are the fulfillment degree, uncertainty degree, possibility degree, and importance degree. The most typical kind of degree is a degree associated to each tuple in a relation (Type 7), for example with the meaning of membership degree of each tuple to the relation, or the fulfillment degree associated to each tuple in the resulting relation after a fuzzy query. Sometimes, it is useful to associate a fuzzy degree to only one attribute (Type 5) or to only a concrete set of attributes (Type 6), for example, in order to measure the truth, the importance, or the vagueness. Finally, in some applications, a fuzzy degree with its own fuzzy meaning (Type 8) is useful in order to measure a fuzzy characteristic of each item in the relation like the danger in a medicine or the brightness of a concrete material.

1.1.2 The FSQL Language

In [33] we can find a good review about the main two extensions to SQL: FSQL and SQLf. The FSQL language [17] is an extension of SQL which allows fuzzy data manipulation like fuzzy queries. FSQL incorporates some definitions to permit the fuzzy information processing:

- Linguistic labels: They are linguistic terms that can be defined on fuzzy attributes. These labels will be preceded with the symbol $ to distinguish them.
- Fuzzy comparators: Besides the typical comparators ($=$, $>$, etc.), FSQL includes fuzzy comparators like "fuzzy equal", "fuzzy greater than", "fuzzy greater or equal", "much greater than", "fuzzy included", etc.
- Function CDEG: The function CDEG (compatibility degree) computes the fulfillment degree of the condition of the query for each answer.
- Fulfillment thresholds: For each simple condition, a fulfillment threshold $\tau \in [0,1]$ may be established (default is 1). Format: $<$ *condition* $>$ THOLD τ indicating that the condition must be satisfied with minimum degree τ to be considered.
- Fuzzy constants: Besides the typical constants (numbers, NULL, etc.), FSQL included many constants such as fuzzy trapezoidal $[a,b,c,d] (see Fig. 1), approximate values using the expression #n (approximately n), fuzzy predefined labels

using $LabelName, crisp intervals with [n,m], UNKNOWN, UNDEFINED, and so forth.

- Fuzzy quantifiers: There are two types: absolute and relative. They allow us to use expressions like "most", "almost all", "many", "very few", etc.

Example 1: "Give me all persons with fair hair (in minimum degree 0.5) that are possibly taller than label $Tall (with a high degree as qualifier)":

```
SELECT * FROM Person
WHERE Hair FEQ $Fair THOLD 0.5
AND Height FGT $Tall THOLD $High;
```

2 From Crisp to Fuzzy: Relational Database Migration

Many organizations aim to integrate the fuzzy information processing in their databases [16]. The best solution is to offer a smooth migration toward this technology. Generally, legacy information system migration allows legacy systems to be moved to new environments that allow information systems to be easily maintained and adapted to new business requirements, while retaining functionality and data of the original legacy systems without having to completely redevelop them [7]. In our context, the migration towards FDBs, or *fuzzy migration*, does not only constitute the adoption of a new technology but also, and especially, the adoption of a new paradigm [4, 6]. Consequently, it constitutes a new culture of development of information systems. In fact, the fuzzy migration of information systems consists in modifying or replacing one or more of their components: database, architecture, interfaces, applications, and so forth, and generally the modification of one of these components can generate modifications of some others. Moreover, migration of data between two systems has more relevance in industrial practice than one would expect at first glance.

Data migration also plays an important role for data warehouse systems. A lot of commercial products are based on database systems to use functionality of the DBMS. Those systems generally require that operative applications are not affected by warehouse queries. Consequently, data is copied into the warehouse [20].

In our previous works [4, 6], we defined this *fuzzy migration* deriving a new FRDB from a RDB, i.e. adapting data, metadata, and the software components accordingly, and all this is made in three phases: Schema migration, Data migration and Programs migration. The definition of this fuzzy migration may involve several problems such as:

- The schemas modification requires a very detailed knowledge on the data organization (data types, constraints, etc.).
- The database source is generally badly documented.
- The difficulty of correspondences establishment between the two databases.
- The database models, source, and target can be incompatible.
- The values in the FMB (metadata) must be chosen after thorough studies.

- The communication protocols between the database and their applications are generally hidden.
- The administrator and at least some database users need some knowledge about fuzzy logic.
- Software using fuzzy information must be designed with care, especially if it will be utilized by regular users.
- Both systems should be run concurrently during enough time.

Few FDBs implementations have been developed in real and running systems [2, 17, 18, 24, 23, 25, 28], and a more recent update about these approaches is in [33, 35]. On the other hand, studies on this kind of migration are scant [4, 6], although there are some general non-fuzzy methods [19, 21, 22, 30, 34].

Basically, our approach [4] is addressed mainly to database administrators (DBA) and consists in including some fuzzy characteristics in any existing database, mainly fuzzy queries. We also study how to optionally migrate the data stored in RDBs towards FRDBs. This approach intend to meet the following requirements:

- to provide for methodical support of the migration process,
- to assist DBA in transforming relational schemas and databases,
- to allow DBA to choose the attributes able to store imprecise data or/and be interrogated with flexible queries,
- to assist DBA in the list of required metadata,
- to exploit the full set of FRDBs features,
- to cover properties of the migration itself such as correctness and completeness.

There are three strategies tailored to the users' needs and we summarize them in the following subsections.

2.1 *Partial Migration*

The goal of this migration is to keep the existing data, schema, and applications. The main benefit in this migration is the flexible querying, but also the FDB may

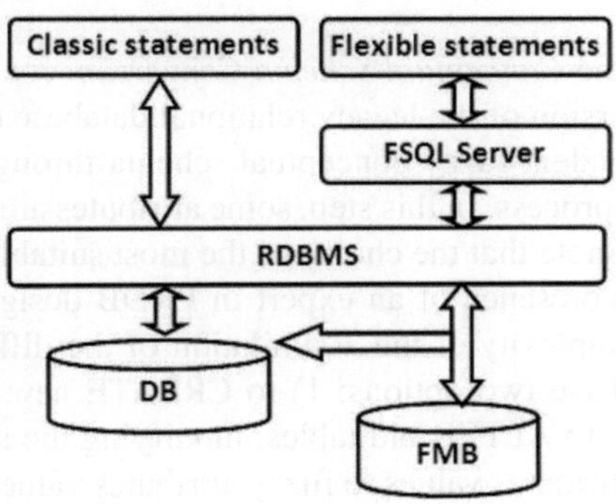

Fig. 2. FRBD Architecture.

use fuzzy degrees, fuzzy quantifiers, and some fuzzy data mining methods could be implemented on crisp data. Here, two elements must be added:

1. The FMB with fuzzy metadata, specially fuzzy attributes, witch must be declared as fuzzy type FTYPE1.
2. The FSQL server to utilize fuzzy statements (queries, deletions, updates, etc.). Fig. 2 shows the DBMS architecture with the FSQL server.

2.2 Total Migration

This strategy offers, in addition to the flexible querying, the possibility to store imprecise data at the level of the fuzzy attributes. Therefore, it will be a total migration towards a FRDB. Contrary to the previous case, now we can use all the fuzzy attributes types: FTYPE1, FTYPE2, FTYPE3, and FTYPE4, and also the degrees (types 5-8). The database modification concerns only the tables defining or referencing these fuzzy attributes. This strategy comprises three main steps: (Step 1) schemas conversion, (Step 2) data conversion, and (Step 3) programs conversion.

2.2.1 Step 1: Database Schemas Conversion

In our context, it consists in modifying the table schemas with fuzzy attributes which are going to store fuzzy values. Moreover, the FMB, which stores the fuzzy attributes information (linguistic labels, similarity relations, etc.), must be created. During this process, the Source Physical Schema (SPS) of the RDB is extracted and then transformed in a correspondent physical schema for the fuzzy DBMS. The new physical schema is used to produce the DDL (Data Definition Language) code of the new FRDB. In this section, we present two strategies of transformations or conversions: *Physical schema conversion* and *Conceptual schema conversion*.

Physical Schema Conversion consists in analyzing the DDL code of the source RDB (stored in the data dictionary) in order to find its physical schema. This relational schema will be converted in the fuzzy DBMS modifying attributes and adding the FMB. The information stored in the FMB for each fuzzy attribute is detailed in [17].

On the other hand, the *Conceptual Schema Conversion* consists in extracting the physical schema conversion of the legacy relational database (SPS) and transforming it into its correspondent fuzzy conceptual schema through a database reverse engineering (DBRE [1]) process. In this step, some attributes are transformed to fuzzy ones. It is necessary to note that the choice of the most suitable fuzzy attribute type is a delicate task. The presence of an expert in FRDB design must be counseled strongly due to the complexity of the assimilation of the different fuzzy concepts [5]. Furthermore, we have two options: 1) to CREATE new tables coping values from old tables, and 2) to ALTER old tables, modifying the structures of these tables, and convert old attributes values to fuzzy attributes values.

[1] A DBRE is a process for studying databases used to recover the conceptual schema that expresses the semantics of the source data structure.

2.2.2 Step 2: Data Conversion

The data conversion consists in transforming the data of the RDB (crisp) to the format of the data defined by the fuzzy schema. Thus, there is a modification in the data representation for fuzzy attributes at the level of the database tables, and this modification could be automated using a defined algorithm. We also must introduce the fuzzy metaknowledge in the FMB (definition of labels, etc.). It should be noted that if we want to "fuzzify" some previous data, then the transformation is not automated. Fuzzy information may be more real than crisp information. As mentioned above, the intervention of an expert in FRDB design and in the database domain is strongly counseled in order to choose the most suitable type among the different types of fuzzy values mentioned previously [5]. Sometimes, the crisp data can be kept. In other cases, they will be transformed, using some standard rules, in linguistic terms, intervals, approximate values, and so forth, Especially in some contexts, NULL values may be transformed into UNKNOWN values.

2.2.3 Step 3: Database Schemas Conversion

The modification of the database structure requires, in the majority of the cases, the modification of their related programs. Legacy code alteration aims at reconciling the application programs with the migrated database. The functionalities of these programs, which now access the renovated database instead of the legacy data, are left unchanged, as well as its programming language and its user interface (they can migrate independently). Section 3 discuss in detail the different strategies and methods used in programs conversion.

2.3 *Easy Total Migration*

We show in the next section that the migration of programs may be a very hard task, but it is mandatory and essential in the total migration. However, the goal of the *Easy Total Migration* strategy is to store imprecise values, to benefit from the flexible querying, fuzzy data mining on fuzzy data, and to keep the existing data and applications with the minimum required modifications. The basic idea is to mix partial and total migration; that is, fuzzy attributes with fuzzy values are duplicated: one with fuzzy values and the other with only crisp values. In this process, we use the three steps of the total migration with some modifications: (Step 1) schemas conversion, (Step 2) data conversion, and (Step 3) programs conversion. In the steps 1 and 2 now we preserve the old attributes. For example, if the Length attribute is a fuzzified attribute converted to FTYPE2, then we preserve the existing Length attribute and add a new attribute for the new fuzzy Length.

The program conversion is now easier, but we must manage the new fuzzy attributes in some DML statements in order to achieve legacy programs running exactly like before the migration:

1. SELECT: No modifications required (except if the SELECT uses the asterisk, *, because it represents all the attributes and in the new FRDB there are more attributes).

2. DELETE: No modifications required.
3. INSERT/UPDATE: The values of the fuzzified attributes must be inserted/updated again in the same row in the corresponding new fuzzy attributes.

The main drawback of this migration strategy is the redundancy in the fuzzified attributes (except the FTYPE1). The main advantage is the easy program migration. In some situations, this is the best option, using this strategy as an intermediate and temporary step to a total migration.

3 Effects of Fuzzy Databases Migration on Programs

A computer application is a whole, made of programs and databases. The data processing rules, nested in the programs, and the methods used to store data in the DB interact with each other to constitute a coherent entity. As we saw above, the modification of the structure of the database requires, in the majority of the cases, propagation to the level of their related programs. In this context, a database migration from a source to a target system should ideally:

- guarantee that the coherence in the interaction "programs database" are kept;
- guarantee that the migrated database will follow the rules imposed by the fuzzy DBMS.

In fact, if we want to use flexible statements with FSQL and store imprecise data, the communication between programs and the database must be through the FSQL server. The programs must be modified according to the representation, interrogation, and storage of the new data. Moreover, we must decide what to do with fuzzy values in each program. Note that emigrating these programs not only means to convert DBMS calls in programs, but in addition requires the reengineering of imperative programs to accept fuzzy values and surely the reconstruction of user interfaces.

Chikofsky [11] defined *re-engineering* as the examination (understanding) and alteration of a system to reconstitute it in a new form and the subsequent implementation of the new form. Therefore *re-engineering* ultimately leads to an almost complete re-implementation of the legacy system (perhaps only a different programming language, or maybe a completely new architecture and technology). The resulting system may or may not run on a different environment.

We thus view *reengineering* as closer to *redevelopment* than to *migration*, which aims to avoid a complete redevelopment of the legacy system. Migration seeks to reuse as much of the legacy information system (implementation, design, specification, requirements) as possible. In addition, the target (fuzzy) system resulting from a migration process runs on a different environment (the FSQL language, the FMB and the FSQL Server), whether it is a different programming language or a completely new architecture and technology. In fact, this new environment is not far from the existing one since it extends it with new features and functionalities. If most of the legacy system must be discarded (no reusable components), the engineer will be facing a redevelopment project, not a migration or maintenance project [1]. Also, although redevelopment involves developing a system from scratch, it requires a thorough understanding of the existing system and thus involves many reengineering activities. In this paper, reengineering is not seen as a solution to the legacy problem per se, but rather as a technology to be used in migration or redevelopment projects.

In fact, we draw our inspiration from the strategies of programs conversion proposed by Henrard et al. [19], Cleve [12, 13] and Jess Bisbal [7]. We propose four methods to adapt the programs of the legacy systems with new fuzzy data: wrapping, maintenance, redevelopment, and rewriting access statements. These methods depend strongly of the three strategies of FDBs migration [4]. Redevelopment involves rewriting existing applications. Wrapping involves developing a software component called wrapper that allows an existing software component to be accessed by other components who need not be aware of its implementation. Migration through rewriting access statements allows legacy systems to be moved to new environments that allow information systems to be easily maintained and adapted to new business requirements, while retaining functionality and data of the original legacy systems without having to completely redevelop them.

3.1 Wrapping Legacy Applications to Support Fuzzy Data

Wrapping techniques provide a natural way of integrating legacy systems with each other and with new software. They constitute a method of encapsulation that provides clients with well-known interfaces for accessing server applications or components. The wrapper layer is the central part of the program conversion phase making correspondence between the relational and the fuzzy relational data format. It simulates the legacy DML[2], such that the logic of the legacy programs does not need to be adapted. The basic principle is to generate for each entity of the source base, all necessary instructions to reproduce the DML verbs (commands) as they were in the source environment. So, for each source entity, generators are created to systematically produce the reading, writing, modifying and deleting instructions, independently from knowing if these instructions are really used by the application. The responsibility of the wrapper layer is twofold:

- structure conversion: the wrappers adapt the legacy database statements to the new fuzzy structure; The resulting data structure should be also adapted to fit to the program requirements.
- language conversion: the wrappers translate the relational DML primitives using fuzzy DML commands;

In their turn, the wrappers which constitute the wrapper layer permit to translate the statements of the legacy system, written in SQL, to the FSQL language (fuzzification in the statement, not in the data). These statements are carried out in the FSQL Server. The FSQL Server translates in its turn these statements to SQL language (defuzzification in the statement) in order to be performed in a relational DBMS[3].

[2] Data Manipulation Language.

[3] The FSQL Server is implemented in Oracle and PostgreSQL DBMS.

These resulting SQL statements use special functions which compute the fuzzy operations. The fuzzy returned answers will be defuzzified in order to be treated by the legacy application programs. In this situation, the application program invokes the wrappers instead of the RDBMS. In other terms, and regarding programs, there must not be a difference between data access in the source relational database and data access in the FRDB. This process is depicted in Fig. 3.

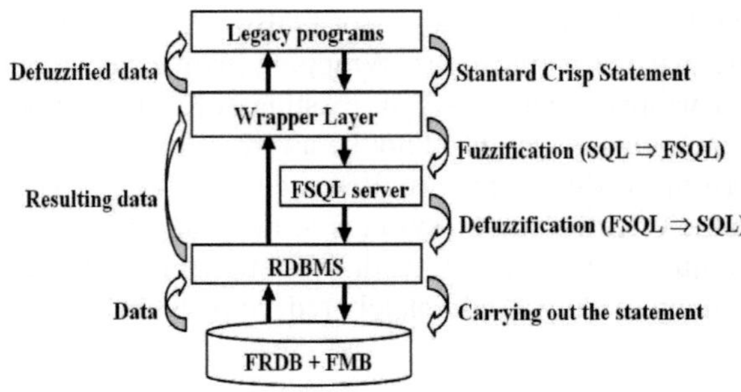

Fig. 3. Programs Migration based on Wrappers.

The advantage of wrapping is that legacy systems (interacted with RDB) become part of the new generation of applications (interacted with FRDB) without discarding the value of the legacy applications. It is a realistic approach, since it is accomplished easily and rapidly with current technology.

The wrapping strategy is used specially in the partial migration. In fact, we revealed that all existing schemas and data of the database are unchanged. For this reason, all existing programs are left unchanged, as well as their programming language and their user interface. However, to benefit of FRDB (flexible querying and some fuzzy data mining methods), in addition to the wrappers implementation, these programs must be maintained in the new system to consider this change. New programs could be developed to reach these fuzzy benefits, instead of modifying legacy programs. These new programs jump the wrapper layer and they are connected with the FSQL Server.

3.2 Redevelopment

From the database standpoint, Redevelopment (rewriting) can be achieved according to two scenarios:

- The database is redesigned as of the previous one did not exist (Fig. 4);
- The designers start from the existing database which they will enrich and modify in order to answer the new users needs (Fig. 5).

In the case of total migration or easy total migration, we revealed that all existing schemas and data of the RDB are converted to the fuzzy schema. In this situation,

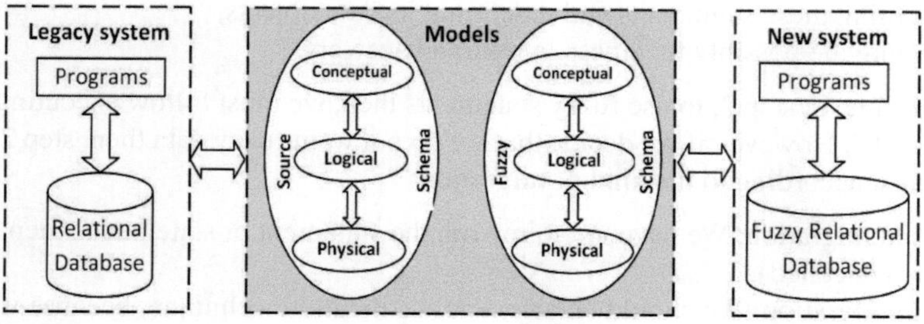

Fig. 4. Redesigning the RDB.

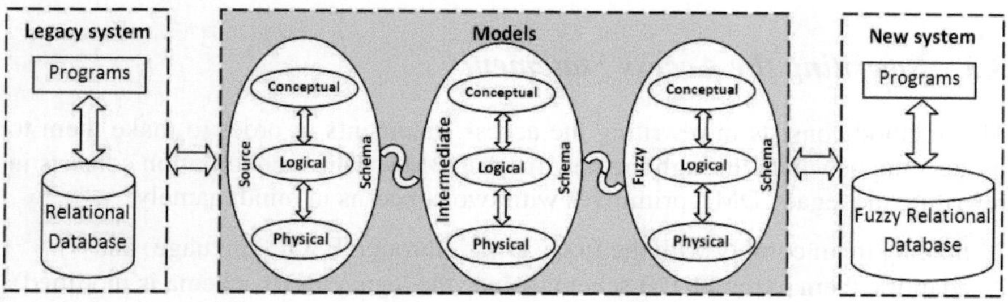

Fig. 5. Fuzzifying the RDB.

the legacy programs are not valid specially in the total migration and, therefore, they must be adapted to the new database format. We have two possibilities:

- Rewriting the access statements
- Redevelopment of these legacy programs.

This method generalizes the problems met in the previous one. In fact, the change of paradigm when moving from standard crisp data in RDB to imprecise ones in FRDB induces problems such as whether the user wants now to use fuzzy information in FSQL statements and the manipulation of the imprecise data returned by these statements. The program is redeveloped (rewritten) in order to use the new fuzzy DBMS-DML at its full power and take advantage of the new data system features. Reengineering is performed here to reduce costs, to increase performance, and to enhance maintainability of the system. The process of reengineering comprises that a system is first analyzed, then changes are made to the system at the abstract level and the system is re-implemented. This strategy is much more complex than the previous one since every part of the program may be influenced by the schema transformation. The most obvious steps consist of:

1. Identifying the statements and the data objects that depend on these access statements,
2. Deciding whether each statement will be now fuzzy or not,

3. Rewriting these statements and redefining its data objects,
4. Treating the possibly fuzziness in returned answers.

If we decide, in step 2, to use fuzzy statements then, we must follow executing both step 3 and 4. However, if we decide that we do not want fuzzy data then, step 2 must be studied according to the kind of migration:

1. Partial migration: We have not to rewrite the statement or statements (step 3 and 4 are cancelled).
2. Total migration: We should choose a defuzzification technique, because we can found fuzzy information in each row.
3. Easy total migration: We have not to rewrite the statement or statements (step 3 and 4 are cancelled), because we can use the non-fuzzy attributes.

3.3 Rewriting the Access Statements

This method consists in rewriting the access statements in order to make them to process the new data through the new fuzzy DBMS. This reconciliation consists in rewriting the legacy DML primitives with two concerns in mind, namely:

1. making them comply with the fuzzy DML (through FSQL language) and
2. adapting them to the FRDB schema (since the legacy RDB schema is modified)

Similarly to the wrapping method, all necessary instructions to reproduce the DML commands are generated as they were in the source environment independently of their use by the application. This is in order to guarantee that all possible maintenance operations could be implemented in the new system. The semantic equivalence that holds between the relational and the fuzzy relational schemas allows the logic of the legacy programs to be left unchanged as well as its programming language and its user interface during the program conversion process. The modification is restricted to access statements. It mainly consists in locally replacing DML primitives with an equivalent code fragment using fuzzy DML primitives. This transformation is based on the mapping that exists between the relational and fuzzy relational database schemas. This task may be complex because programs will manage imprecise data instead of legacy crisp ones. The more flexible option includes the possibility for the user of setting all the required parameters for the DML statements (fuzzy labels, fuzzy constants and specially fuzzy thresholds). The results of a fuzzy query always have a fulfillment degree (function CDEG in FSQL), we can forget it, but we can also allow the user to dynamically change the threshold, in order to get more or less tuples.

3.4 Maintenance of Legacy Programs

The use of FRDBs is justified essentially by the advantages of fuzzy queries, fuzzy data mining and the storage of fuzzy data. For this reason, the migration toward FRDBs must be followed by an updating of the programs using them. This updating includes maintenance, rewriting access statements or redevelopment. Concerning the maintenance, we can say that it cannot be effectively a method of updating programs because it can be done at different time of the life-cycle of any program. The programs must be maintained in the new system to consider the FRDB features. Maintenance has been added for completeness in order to improve the user interface by adding the novelties introduced by FRDBs. Maintenance can be applied to the partial migration. For example, while consulting the FMB, the user interface can provide a list of linguistic terms, fuzzy quantifiers, satisfaction degrees, etc. It could consider the answers returned by the FSQL Server, their degree, their sorting, some data mining operations, etc. With regard to the total migration, all programs using the new fuzzy attributes need modifications, except probably attributes FTYPE1. We can distinguish two kind of modifications:

- Fuzzy information processing: The knowledge engineers should think and decide the required operations on each fuzzy attribute.
- Input/Output of fuzzy information: The interfaces should be modified to cope with the new kind of information. We can use graphical interfaces (drawing the fuzzy values, ...), menus with fuzzy labels, tools for defining new fuzzy values and fuzzy labels according with real and/or expert data, etc.

4 Example

Let FLAT be the database relation described in Table 1. Let us suppose that we have a program which interrogates this table to find "the cheap flats near to Tunis Center". This query can be written in SQL in the following way:

Table 1. Extension of the table FLAT

FlatNumber	Address	Rent	Distance_to_Tunis
1	Bardo	320	2
2	Ariana	290	11
3	La Marsa	400	15
4	Nabeul	200	40
5	Tunis	310	0

```
SELECT * FROM FLAT
WHERE Distance_to_Tunis < 10
AND Rent BETWEEN 150 AND 300;
```

In the legacy system, this query could return empty answers. The flats numbers 1 and 5 satisfy the first query criterion (`Distance_to_Tunis`) but they are slightly away from the second criterion (`Rent`). However, the flat number 2 isn't appear in the result because it is slightly away from Tunis (11 km $\approx$ 10 km). Reasonably, if we inform the user about the real situation, he/she could accept one of the flats of number 1, 2 and 5.

In the new system, the user must be more satisfied. If we are in the situation of rewriting statement access, the application programs are modified such as they provide a list of linguistic terms to the user in order to express its needs better. The user interface is updated and consulting the FMB, it is able to present the following terms: Low, Medium, Near, Far, High, Cheap, Large, etc. It can provide also a satisfaction degree (0% → 100% or in the [0,1] interval). Selecting these terms and specifying a degree, the application reformulates the legacy SQL query. The new FSQL query may be written now as follow:

```
SELECT % FROM FLAT
WHERE Distance_to_Tunis FEQ $Near 0.8
AND Rent FEQ $Cheap 0.5;
```

This query is transmitted to the FSQL Server and now we can get always the more interesting tuples since they cover the possible nearest (neighbor) answers. Furthermore, if we get an empty answer we always can to reduce the thresholds. On the other hand, if we increase the thresholds, we can get fewer tuples.

However, when legacy programs codes are inaccessible or if we want to keep all the programs unchanged and benefit of flexible queries (using other than legacy programs), we have to follow wrapping strategy. The SQL queries are translated simply, at the level of the wrapper layer, to FSQL queries with a fixed satisfaction degree (for example 0). This wrapper layer, captures results from the FRDB, converts them to the appropriate legacy format and delivers them to the application program. In this situation, the answers could be very large and less precise because the satisfaction degree is not chosen by the user.

5 Conclusions and Futures Lines

Legacy Information System Migration studies how to move legacy systems to new environments in order to allow information systems to be easily maintained and adapted to new business requirements, while retaining functionality and data of the original legacy systems without having to completely redevelop them. In previous works [4], we showed three legacy information system migration strategies, from classic RDBs to fuzzy relational ones. These strategies allow us to reach the advantages of FRDBs: fuzzy queries, fuzzy data mining [15], storage of fuzzy information and, in general, fuzzy information processing [16]. Probably the most interesting tool are the fuzzy queries [35], because, for example, they allow more precise queries, rank the results and tune the query to get the desired number or tuples.

Thus, it is not very strange that, many organizations aim to integrate the FRDBs advantages, minimizing the transformation costs. This chapter summarized the main strategies to this goal about the migration towards FRDBs. We concentrate our efforts in the more complex task in this kind of migration: The migration of programs. In this line, we propose four methods to adapt the programs of the legacy systems with new fuzzy data: wrapping, maintenance, redevelopment, and rewriting access statements. These methods are related strongly to the strategies of FRDB migration presented in our previous work [4]. Redevelopment involves rewriting existing applications. Wrapping involves developing a software component called wrapper that allows an existing software component to be accessed by other components who need not be aware of its implementation. Migration through rewriting access statements allows to process the new data through the new fuzzy DBMS while modifying only the statements which access to the database.

Future work will focus on the further enhancement and development of the presented methods. Among others, the incorporation of FRDBs access Application Programming Interface (API). In fact, this API helps to: ensure security, insulate the application from having to deal directly with the database server, and provide query services for multiple types of FRDB products. Computer programs written in Visual Basic, C, C++, Pascal, COBOL and many other languages could perform FRDB operations via ODBC[4]. Similarly, programs written in Java could use JDBC[5] to perform FRDB operations. We plan on defining ODBC and JDBC drivers for FRDBS. In this situation, programs using ODBC or JDBC can run directly on the FRDB server computer (FSQL Server + DBMS), but, more typically, they run on client computers networked to a database server.

Acknowledgements. This work has been partially supported by the "Ministry of Education and Science" of Spain (projects TIN2006-14285 and TIN2006-07262) and the Spanish "Consejería de Innovación Ciencia y Empresa de Andalucía" under research project TIC-1570.

References

1. Alan, R., Simon, S.: Systems Migration: A Complete Reference. Van Nostrand Reinhold, New York (1993)
2. Barranco, C., Campaña, J., Medina, J.: Towards a fuzzy object-relational database model. In: Galindo, J. (ed.) Handbook of Research on Fuzzy Information Processing in Databases, pp. 435–461. Information Science Reference, Hershey (2008), http://www.info-sci-ref.com
3. Bellman, R., Zadeh, L.: Decision-making in a fuzzy environment. Manage Sciences 17(4), 141–175 (1970)
4. Ben Hassine, M.A., Grissa, A., Galindo, J., Ounelli, H.: How to achieve fuzzy relational databases managing fuzzy data and metadata. In: Galindo, J. (ed.) Handbook on Fuzzy Information Processing in Databases, pp. 351–380. Information Science Reference, Hershey (2008), http://www.info-sci-ref.com
5. Ben Hassine, M.A., Grissa-Touzi, A., Ounelli, H.: About the choice of data type in a fuzzy relational database. In: Gupta, B. (ed.) Computers and Their Applications, pp. 231–238. ISCA (2007)

[4] Open Database Connectivity is a multidatabase API for accessing a database.

[5] Java Database Connectivity is a Java API for connecting programs written in Java to the data in relational databases.

6. Ben Hassine, M.A., Ounelli, H., Touzi, A.G., Galindo, J.: A migration approach from crisp databases to fuzzy databases. In: Proc. IEEE International Fuzzy Systems Conference FUZZ-IEEE 2007, London, pp. 1872–1879 (2007), doi:10.1109/FUZZY.2007.4295651
7. Bisbal, J., Lawless, D., Wu, B., Grimson, J.: Legacy information systems: Issues and directions. IEEE Software 16(5), 103–111 (1999), `citeseer.ist.psu.edu/article/bisbal99legacy.html`
8. Bosc, P.: Fuzzy databases. In: Bezdek, J.C. (ed.) Fuzzy Sets In Approximate Reasoning And Information Systems, pp. 403–468. Kluwer Academic Publishers, Dordrecht (1999)
9. Bosc, P., Ludovic, L., Pivert, O.: Bases de données et flexibilité: les requêtes graduelles. Technique et science informatiques 17(3), 355–378 (1998)
10. Buckles, B.P., Petry, F.E.: A fuzzy representation of data for relational databases. Fuzzy Sets and Systems 7, 213–221 (1982)
11. Chikofsky, E.J., Cross II, J.H.: Reverse engineering and design recovery: A taxonomy. IEEE Softw. 7(1), 13–17 (1990), `http://dx.doi.org/10.1109/52.43044`
12. Cleve, A., Henrard, J., Hainaut, J.L.: Co-transformations in information system reengineering. Electr. Notes Theor. Comput. Sci. 137(3), 5–15 (2005)
13. Cleve, A., Henrard, J., Roland, D., Hainaut, J.L.: Wrapper-based system evolution application to codasyl to relational migration. In: 12th European Conference on Software Maintenance and Reengineering, CSMR 2008, Athens, Greece, pp. 13–22. IEEE, Los Alamitos (2008), `http://dblp.uni-trier.de/db/conf/csmr/csmr2008.html`
14. Elmasri, R., Navathe, S.B.: Fundamentals of Database Systems, 4th edn. Addison-Wesley Longman Publishing Co., Inc., Boston (2003)
15. Feil, B., Abonyi, J.: Introduction to fuzzy data mining methods. In: Galindo, J. (ed.) Handbook of Research on Fuzzy Information Processing in Databases, pp. 55–95. Information Science Reference, Hershey (2008), `http://www.info-sci-ref.com`
16. Galindo, J. (ed.): Handbook of Research on Fuzzy Information Processing in Databases. Information Science Reference, Hershey (2008), `http://www.info-sci-ref.com`
17. Galindo, J., Urrutia, A., Piattini, M.: Fuzzy Databases: Modeling, Design, and Implementation. IGI Publishing, Hershey (2006)
18. Goncalves, M., Tineo, L.: SQLf vs. Skyline: Expressivity and performance, Vancouver, Canada, pp. 2062–2067 (2006)
19. Henrard, J., Hick, J.M., Thiran, P., Hainaut, J.L.: Strategies for data reengineering. In: WCRE 2002: Proceedings of the Ninth Working Conference on Reverse Engineering (WCRE 2002), pp. 211–220. IEEE Computer Society, Washington (2002)
20. Immon, W.H.: Building the Data Warehouse, 4th edn. John Wiley & Sons, Inc., Chichester (2005)
21. Jahnke, J., Wadsack, J.: Varlet: Human-centered tool support for database reengineering (extended abstract). In: Proceedings of Workshop on Software-Reengineering, WCRE 1999 (1999), `citeseer.ist.psu.edu/jahnke99varlet.html`
22. Jeusfeld, M.A., Johnen, U.A.: An executable meta model for re-engineering of database schemas. In: Loucopoulos, P. (ed.) ER 1994. LNCS, vol. 881, pp. 533–547. Springer, Heidelberg (1994)
23. Kacprzyk, J., Zadrożny, S.: On a fuzzy querying and data mining interface. Kybernetika 36(6), 657–670 (2000)
24. Kacprzyk, J., Zadrozny, S.: Fuzzy querying for microsoft access. In: Proceedings of Third IEEE International Conference on Fuzzy Systems - FUZZ-IEEE 1994, vol. 1, pp. 167–171 (1994)

25. Kacprzyk, J., Zadrozny, S.: Computing with words in intelligent database querying: standalone and internet-based applications. Inf. Sci. 134(1-4), 71–109 (2001)
26. Medina, J.M., Pons, O., Vila, M.A.: GEFRED: A generalized model of fuzzy relational databases. Information Sciences 76(1-2), 87–109 (1994), citeseer.ist.psu.edu/medina94gefred.html
27. Prade, H., Testemale, C.: Fuzzy relational databases: Representational issues and reduction using similarity measures. Journal of the American Society for Information Science 38(2), 118 (1987)
28. Tineo Rodríguez, L.J.: Extending RDBMS for allowing fuzzy quantified queries. In: Ibrahim, M., Küng, J., Revell, N. (eds.) DEXA 2000. LNCS, vol. 1873, pp. 407–416. Springer, Heidelberg (2000)
29. Takahashi, Y.: Fuzzy database query languages and their relational completeness theorem. IEEE Trans. on Knowl. and Data Eng. 5(1), 122–125 (1993)
30. Tilley, S., Smith, D.B.: Perspectives on legacy systems reengineering. Software Engineering Institute, Reengineering Center (Carnegie Mellon University) (1995), http://www.sei.cmu.edu/reengineering/lsysree.pdf
31. Umano, M.: Freedom-o: A fuzzy database system. Fuzzy Information and Decision Processes, 339–347 (1982)
32. Umano, M., Fukami, S.: Fuzzy relational algebra for possibility-distribution-fuzzy-relational model of fuzzy data. J. Intell. Inf. Syst. 3(1), 7–27 (1994)
33. Urrutia, A., Tineo, L., Gonzalez, C.: FSQL and SQLf: Towards a standard in fuzzy databases. In: Galindo, J. (ed.) Handbook on Fuzzy Information Processing in Databases, pp. 270–298. Information Science Reference, Hershey (2008), http://www.info-sci-ref.com
34. Wu, B., Lawless, D., Bisbal, J., Richardson, R., Grimson, J., Wade, V., O'Sullivan, D.: The butterfly methodology: A gateway-free approach for migrating legacy information systems. In: ICECCS 1997: Proceedings of the Third IEEE International Conference on Engineering of Complex Computer Systems (ICECCS 1997), p. 200. IEEE Computer Society, Washington (1997)
35. Zadrozny, S., Tré, G.D., de Caluwe, R., Kacprzyk, J.: An overview of fuzzy approaches to flexible database querying. In: Galindo, J. (ed.) Handbook on Fuzzy Information Processing in Databases, pp. 34–54. Information Science Reference, Hershey (2008), http://www.info-sci-ref.com
36. Zemankova, M., Kandel, A.: Implementing imprecision in information systems. Inf. Sci. 37(1-3), 107–141 (1985)

Author Index

Atanassov, Krassimir 61

Ben Hassine, Mohamed Ali 175
Bordogna, Gloria 79
Bosc, Patrick 133
Bourgeois, Brian S. 3

Chountas, Panagiotis 61

De Tré, Guy 9
Dujmović, Jozo 9

Galindo, José 175

Iyengar, S.S. 155

Jankowski, Piotr 3

Kacprzyk, Janusz 117

Laurent, Anne 43

Morris, Ashley 3

Oğuztüzün, Halit 97
Ounelli, Habib 175

Pagani, Marco 79
Pasi, Gabriella 79
Petry, Frederick E. 3, 97
Pivert, Olivier 133

Robinson, Vincent B. 29
Rogova, Ermir 61

Sözer, Aziz 97

Van de Weghe, Nico 9
Vert, Gregory 155

Yazıcı, Adnan 97

Zadrożny, Sławomir 117

MIX
Papier aus verantwortungsvollen Quellen
Paper from responsible sources
FSC® C105338

If you have any concerns about our products,
you can contact us on
ProductSafety@springernature.com

In case Publisher is established outside the EU,
the EU authorized representative is:
Springer Nature Customer Service Center GmbH
Europaplatz 3, 69115 Heidelberg, Germany

Printed by Libri Plureos GmbH
in Hamburg, Germany